P9-DDV-097

WHY DIETS MAKE US FAT

CURRENT

WHY DIETS MAKE US FAT

The Unintended Consequences of Our Obsession with Weight Loss

Sandra Aamodt, Ph.D.

Current

CURRENT
An imprint of Penguin Random House LLC
375 Hudson Street
New York, New York 10014
penguin.com

LIBRARY OF CONGRESS CATALOGING-IN-PUBLICATION DATA
Names: Aamodt, Sandra.
Title: Why diets make us fat : the unintended consequences of our obsession with weight loss / Sandra Aamodt, Ph.D.
Description: New York, New York : Current, an imprint of Penguin Random House LLC, [2016] | Includes bibliographical references and index. | Description based on print version record and CIP data provided by publisher; resource not viewed.
Identifiers: LCCN 2015048708 (print) | LCCN 2015048518 (ebook) | ISBN 9780698186668 (ebook) | ISBN 9781591847694 (hardcover : alk. paper)
Subjects: LCSH: Weight loss—Psychological aspects | Weight gain. | Food habits—Psychological aspects
Classification: LCC RM222.2 (print) | LCC RM222.2 .A225 2016 (ebook) | DDC 613.2/5—dc23
LC record available at http://lccn.loc.gov/2015048708

Printed in the United States of America
1 3 5 7 9 10 8 6 4 2

Set in ITC Galliard Std
Designed by Alissa Rose Theodor

For Elizabeth

Contents

WHY DIETS MAKE US FAT

Introduction

Several years ago, my husband and I visited friends who had two teenage daughters. Both girls were smart, beautiful, and well loved, but only one was thin. Her sister had started gaining weight at puberty, and their parents couldn't figure out what to do. They'd tried talking to her about portion control and exercise. They'd tried cooking her special food. Nothing slowed her progression from skinny child to voluptuous woman.

Their well-meaning concern was making the situation worse. She was reluctant to sit down for meals with the family—and ate almost nothing when she did. But her father told me that food was disappearing from the refrigerator in the middle of the night, and her weight gain continued. Rather than making her thin, her parents' anxiety was pushing her toward an eating disorder and increasing her odds of gaining more weight.

I understood her feelings because I was that teenage girl myself several decades ago, hating myself because I couldn't satisfy my culture's expectations for my body. Because I grew up to be a neuroscientist, I also knew about the biology behind her struggles. Considerable evidence shows that dieting—willfully eating less to lose weight—rarely works in the long run, while its unintended side effects do a lot of harm.

When I discovered this evidence, after three decades of losing the same fifteen pounds again and again, it transformed my relationship with food. At first I wasn't sure that changing my approach would work for me, so I tried an experiment. As my New Year's resolution in 2010, I vowed to go an entire year without dieting or weighing myself and to exercise every day. I was so pleased with the results that I've maintained all three habits ever since. My weight used to bounce around a fifty-pound range as I starved or binged my way through the first half of my adult life, but now it remains stable even though I don't worry about calories anymore. And the research suggests that my experience is typical.

If diets worked, we'd all be thin by now. Instead we have enlisted hundreds of millions of people into a war we can't win, fighting not against an external enemy but against our own bodies. In the United States, 108 million people went on a diet last year—and half of them were within the normal weight range when they started. Yet research shows that dieters almost always regain their lost weight within a few years, no matter which diet plan they follow. In fact, they often regain more weight than they've lost. On average, people who go on a diet end up heavier five years later than people of the same initial weight who didn't diet.

Many people are convinced that unless they control their appetite, they will gain weight, get sick, and die young. This common belief gets it backward. Dieting often makes people fatter and less healthy in the long run because it impairs their ability to recognize hunger, which increases vulnerability to emotional eating and food marketing. Research suggests that the resulting cycle of repeatedly losing weight and gaining it back is no better for health than maintaining a higher weight with a healthy lifestyle. Fitness can overcome most of the health risks associated with obesity and also help prevent future weight gain, but only if people don't quit exercising when they discover that it isn't making them lose weight.

By understanding how the brain regulates weight, we can stop struggling to defeat this control system and learn how to let it help us instead. Just as the body needs a certain amount of sleep, the brain has a body weight range that it prefers and will defend for each individual. The brain's weight-regulation system will maintain a stable, healthy weight for most people, if it is allowed to do its job without interference. This system mainly operates beneath conscious awareness, influencing hunger and activity in many ways and also changing metabolism by modifying the amount of energy that muscles burn. Lasting weight loss is hard to achieve because when dieting makes a heavy person thinner, the brain responds exactly as it would if a thin person were starving. The result is a powerful mechanism to keep body weight within a certain range.

The belief that dieting will control weight is not only mistaken but also dangerous. I wrote this book because I've seen too many healthy children turn into diet-obsessed teenagers who are headed for a lifetime of weight cycling and struggles with food. Parents often unwittingly contribute to these problems by hassling kids about their weight or by setting an example of constant anxiety around meals. With the help of a good therapist, our friends' daughter found a better way to relate to food. I want us all to do that, for our children and for ourselves.

Our obsession with weight loss hasn't made us thin or healthy, so it's time to try another way. If dieting has failed you, that's not because you're a failure but because your brain is working correctly. To find out what to do instead, read on.

Part One

The Trouble with Diets

1

The Diet Roller Coaster

His oldest daughter's wedding was the final straw. Dennis Asbury could avoid looking in mirrors, but he couldn't ask his wife not to hang photos from the celebration on the wall. The truth was staring at him: at age forty-five, he carried 283 pounds on his 5'9" frame and had a fifty-two-inch waist. Stress at his job as director of environmental services for the town of Eden, North Carolina, had led him to overeat for years. Now the bill had come due. The next week at a checkup, he asked his doctor for help with weight loss.

Like six million other people in the United States, he took home a prescription for fenfluramine-phentermine, better known as fen-phen.[1] The drug combination was everywhere in 1996, after a strong marketing push from its manufacturer, American Home Products, based on a four-year study of only 121 patients. For Dennis, the drugs were a revelation. After taking one pill, he felt like he was in someone else's body. For the first time in years, he could get through an entire thought at work without anticipating his next meal. Encouraged, he plunged into his diet with enthusiasm.

As a numbers guy, Dennis approached his new lifestyle with the spirit of an engineer, logging every bite of food and each minute of exercise. If he did the math correctly, he believed, everything else would work out, whether the problem was a city budget or a calorie budget. He gave up all the foods he loved and tried to move as much as possible, often waking at three a.m. to walk the streets. "Fortunately, I worked for the city," he says, "and most of the police officers knew me." In the evening, he walked at least another hour and then lifted weights. Instead of following a specific diet, he tried to keep his calorie intake as low as possible, sometimes managing to hit six hundred calories for several days in a row. In six months, he had reached 160 pounds. He was thrilled, and so were his wife and his doctor, who took him off the drugs in the spring of 1997. The hard part was over, Dennis thought, although he planned to keep tracking his calories and exercise for life to avoid regaining the weight.

By losing so much weight through diet and exercise, Dennis had already beaten long odds. Men in his starting weight range have only a 1-in-1,290 chance per year of reaching the normal range.[2] Yet keeping the weight off proved even more difficult, as the research predicts. Abundant scientific evidence indicates that almost everyone who loses a lot of weight ends up gaining it back in the long run. That's not because dieters are lazy or lack willpower. Instead the culprit is biology: in particular, the brain's ability to fight back against weight loss.

The Defended Weight Range

The brain has powerful ways of defending the body weight that is its target. This target, often called the set point, is really a range, spanning about ten or fifteen pounds. Within this range, weight is relatively easy to modify through lifestyle changes, while staying below the brain's target is difficult. Over time, the defended range can change—but it goes up much more easily than it goes down. This imbalance means that weight gain is often not reversible. People are likely to be thinner in the long run if they focus on keeping their brain's defended weight range from increasing in the first place than if they rely on losing weight after they've kept it on for a while.

The brain's weight-regulation system has two simple priorities: body weight should remain stable and shouldn't decrease too much. Both these goals were sensible for most of human history (starvation is a much worse fate than obesity), but in modern life, with high-calorie food so easy to find, the second priority has become a problem for many people. Efforts to defeat the brain's resistance to weight loss often end up increasing the influence of external cues like advertising that promote overeating.

When people are obese, it's usually because their brain considers that weight to be correct for them. If someone started at 120 pounds and dieted down to 80 pounds, her brain would declare a starvation state of emergency, using every method available to get her weight back up to normal—and exactly the same thing would happen to someone who started at 300 pounds and lost down to 200 pounds. Although we often assume that the brain should know the difference between a normal-weight person and an obese person when it comes to weight loss, considerable evidence indicates that the brain doesn't distinguish the two situations.

All experts agree that the long-term success rate for weight loss is bad; the only question is how bad. Optimistic researchers suggest that 80 percent of people who lose weight on purpose gain it back over the next few years, while others think the rate is closer to 100 percent. The exact number depends on how success is defined: how much weight, kept off for how long. By the most forgiving definition—10 percent of the lifetime highest (nonpregnant) body weight kept off for at least a year at the time of the survey—*Consumer Reports* found about a quarter of overweight people to be successful losers. Using the same definition, another survey[3] found that 21 percent of obese people qualified.

By that standard, I'm a diet success story myself. In 2008, my husband and I took off from San Francisco to cross the ocean in a forty-five-foot sailboat. On the passage it was difficult to exercise, and fresh food was scarce and expensive on many of the South Pacific islands that we visited. Combined with the stress and sleep deprivation (because one of us needed to keep watch all night at sea), those conditions formed a perfect storm for weight gain. When we reached New Zealand five months later, I'd put on about twenty pounds.

Once we were back in reach of walking trails and farmers' markets, I dieted for several months to get back to my normal weight. Since then, I've kept the extra pounds off without making any special effort—because that higher weight was above my defended range, and so my brain is not trying to push me back to it. In general, people who temporarily go above their normal weight range, for instance during an episode of depression, find it easier to lose the weight and keep it off, especially if they haven't stayed at the higher weight for too long.

Most people's defended range is near their current weight, though it may be either higher or lower depending on their habits. People who binge or often eat when they aren't hungry may be maintaining a weight above their defended range. People who often find themselves hungry or cold are probably below it. The way to discover your own defended range is to give up food restriction, eating only when you're hungry and stopping as soon as you're full. Within six months to a year, your body weight should stabilize within your brain's desired range.

Saying that it's not possible to lower the defended range by changing diet and exercise habits isn't the same as saying that what people eat has no influence on their weight. Of course it does, especially in the short term. It's easy to reduce the temperature in a house by leaving a window open on a cold day, but that doesn't affect the setting on the thermostat, which is going to pump extra heat through the vents to compensate for the temperature drop. The brain's energy-balance system works the same way, responding to dieters' actions by using powerful tools to push their weight back to its starting value.

The weight "thermostat" is located in a brain area called the hypothalamus. This area contains neurons that regulate a bunch of important basic functions, from sex to sleep to body temperature to eating and drinking, mostly without conscious awareness. The part of the hypothalamus that regulates energy balance receives a variety of signals from the body about fat stores, blood sugar, nutrients, food intake, and so on. In response, it modifies hunger, activity, and metabolism to keep body weight within its defended range.

Nothing tastes as good as thin feels, or so the motivational speakers tell us. How does it feel to be substantially below the defended range? Most people feel hungry, cold, and sluggish. With a lower metabolism, they burn less energy, which means that they produce less body heat. The hypothalamus also conserves energy by reducing unconscious motion, like jiggling the leg while sitting—another way people keep themselves warm. In short, if the hypothalamus isn't happy, the body isn't happy.

My husband has never forgotten the intensity of that feeling, although he experienced it more than thirty years ago. On a solo hiking trip in the Alaskan wilderness during the summer after he finished college, he miscalculated how much food he would need. When preparing for the trip, he didn't account for the extra energy costs of walking across tundra, which is harder than trail walking, or of maintaining his body temperature in a cold environment. He lost ten pounds over three weeks. Hunger was a constant companion on the hike, and when he got back to civilization where he could eat as much as he wanted, he still couldn't stop thinking about food. He kept pace at the table with construction workers who did hard labor all day, ate lots of grilled-cheese sandwiches and peanut butter straight from the jar, and continued to stuff himself when his stomach hurt from all the food he'd eaten. Once his weight returned to its normal range, his urge to binge disappeared.

A One-Way Ratchet

Modern life offers many opportunities for the defended range to inch upward. As puberty changes the proportion and location of body fat, it can cause lasting weight gain and much adolescent angst. Staying above the current weight range for too long, for any reason, can lead the brain to begin to defend the higher weight. How long is too long varies from person to person, but typically it's a matter of years. Perhaps as a special case of this rule, pregnancy can permanently increase weight in some women. A temporary weight gain is easier to reverse soon after it occurs, before the brain decides that the situation is the new normal. People who have already raised their defended range, though, can still enjoy a long and healthy life, as we'll see in part 3.

The defended weight range also tends to creep up with age, by about a pound a year, especially in people who don't exercise enough. That adds up to a lot of weight over decades, but good lifestyle choices can reduce or prevent this age-related gain. One lifestyle choice that predicts weight gain is eating a typical Western diet.[4,5] In long-term studies, people who eat more processed meat, more sugar, fewer whole grains, and fewer fruits and vegetables gain more weight than people who have the opposite dietary habits.

As we might expect, access to yummy food also promotes weight gain in laboratory animals and in people.[6] Rodents that are fed high-fat chow weigh 10 or 20 percent more than siblings who eat normal chow, on average, especially if the tasty diet is started early in life. (A rodent on a high-fat diet in a laboratory experiment may get as much as 70 percent of its calories from fat, compared with an average of 35 to 40 percent for people in the United States.) Just as in people, results vary across individual rodents, depending on genetic type. Some animals gain weight simply by being allowed to eat as much regular chow as they want. Other animals eat no more calories on the high-fat diet than they do on their regular diet, like that annoying person who appears to require no effort to resist the doughnuts in the lunchroom at work. As researchers increase the calorie content of the available food, though, most animals eventually gain weight. This so-called diet-induced obesity is probably the best model for the increase in human obesity over the past few decades.

Contrary to popular belief, the process does not seem to be reversible. Keeping body weight below its defended range, even for years, does not lead the brain to regard the thinner state as its new aim.[7] This persistence is the fundamental reason that it is so difficult to maintain a weight loss—because the brain continues trying to push weight back up long after dieters have achieved their goal and moved on with their lives. That's why it's not wise to try to lose weight by making any lifestyle changes that can't be maintained forever. Any other approach is nearly guaranteed to result in weight cycling (yo-yo dieting).

The few known ways to decrease the defended range may have the unwanted side effect of killing people. Cachexia is a state of malnutrition resulting from illness, in which patients have little appetite and a racing metabolism.[8] This disturbance, seen in patients with cancer, AIDS, Alzheimer's disease, and many other disorders, is probably caused by inflammation-related molecules acting on the brain's weight-regulation system to lower the defended range. It tends to occur at the end stages of illness, is difficult to treat, and often leads to death—which doesn't make it promising as a sustainable weight-loss approach.

A lifestyle change that would reduce most people's weight is to take up smoking. Nicotine makes people less hungry and increases the energy that they burn by revving up their metabolism. That's why people who quit smoking tend to gain some weight.[9] Researchers estimate that the decline in smoking accounts for about a quarter of the recent rise in obesity among men and one sixth of the rise among women.[10] For anyone worried about their health, though, smoking is much worse than carrying a few extra pounds.

Many people believe that the success of weight-loss surgery must prove that eating less can lead to permanent weight loss if dieters stick with the program. Surprisingly, such surgery's lasting effects are also due to its influence on the brain's defended weight range. We don't yet know exactly how the gut communicates to the brain that now would be a good time to drop some weight, but it's clear that stomach capacity isn't the answer.

That's easiest to show in rodents because researchers can monitor eating and activity precisely in the laboratory. In one study, for the first three weeks after surgery, rats ate less than they had before, losing about 20 percent of their body weight.[11] After that, they ate their usual number of calories (in more frequent, smaller meals) but did not regain the weight. When the same rats later lost extra weight because researchers put them on a low-calorie diet, the rats ate more until they had returned to their usual postsurgical weight. Because the rats could eat enough to gain weight, but instead maintained a new, lower weight after the operation, the researchers concluded that the surgery had decreased the defended range.

Similarly, people who undergo bariatric surgery tend to lose a lot of weight in the first year or two and then gain some of it back, showing that they too can eat enough to gain weight.[12] Women who've had the surgery can also gain weight during pregnancy and then lose it again. Patients' brains have adjusted to a new defended range, though, as their weight typically remains stable from five to at least fifteen years after surgery, at 20 to 30 percent below its starting value.[13] Many patients report immediate decreases in hunger after surgery.

Despite its effectiveness, bariatric surgery is also on the list of things that can kill people or permanently ruin their health. The risk of dying due to the operation is roughly one in three hundred.[14] For every hundred people who have the surgery, seven to twenty return to the hospital because of postsurgical complications.[15,16] Most patients experience at least one side effect of surgery, with malnutrition and gallstones being the most common. Such serious risks are justified only for people who face disabling health problems, as weight-related illness predicts who will benefit from bariatric surgery much better than weight alone.[17]

Despite these dangers, unscrupulous clinics target women looking to improve their appearance. The Food and Drug Administration (FDA) warned two California companies to remove billboards that advertise the operation without mentioning its risks.[18] If health concerns are the main reason for this surgery, it's hard to explain why 84 percent of the patients who had it done in 2002 were female.[19] Among people who meet the medical criteria for bariatric surgery, women are three times more likely than men to have the operation. Their willingness to take such risks underscores the cultural pressure on women to be thin.

What about diet drugs like the ones that helped Dennis Asbury? Fen-phen isn't available anymore. In September 1997, the drug combination was taken off the market after the FDA concluded that fenfluramine causes a serious side effect. The drug damaged heart valves in 23 percent of people who took it, leading to hundreds of deaths.[20] Faced with a class-action lawsuit, the drug company Wyeth (successor to American Home Products) set aside $21.1 billion for people who had been harmed.[21] Dennis's heart was tested and pronounced healthy, so he was puzzled when he received a small check from the drug company as part of the settlement. As far as he was concerned, the drugs had done their job, but for many other dieters that prescription was a disaster.

In 2012, the FDA approved the first new weight-loss drugs in thirteen years. The decision followed a $60 million lobbying campaign by a consortium of weight-loss clinics, physicians, and drug companies begun in 2010.[22] In 2011, the Senate Agriculture Committee, which controls FDA funding, directed the agency to report what it was doing to provide new drugs for obesity. Over the next three years, five new drugs were approved. The FDA had previously rejected three of them (Belviq, Qsymia, and Contrave) because of concerns that their risks outweighed their benefits.

Prescription drugs for weight loss work only in combination with calorie restriction and exercise. They have modest effects, causing people to lose five to fifteen pounds more, on average, than they would have on the diet alone. None of these drugs changes the defended weight range, so once someone stops taking the drug, the lost weight is likely to come back within a year. Another new option, an implanted device that stimulates the vagal nerve to reduce appetite, was approved in January 2015. In its most successful clinical trial, people using the device lost an average of 6.7 pounds more than those on a diet and exercise program alone after eighteen months.[23] That seems like a small reward for having a device implanted in the body, with all the associated risks.

Diets Fail in the Long Run

People who want to maintain a weight below their defended range have taken on a lifelong project. If dieting could lower the defended range, we might expect that maintaining a lower weight would become automatic eventually. But it doesn't seem to work that way. Most people continue to struggle even after maintaining a weight loss for a long time. In one study, among people who had already kept off thirty pounds or more for at least a year, more than a third gained weight (an average of fifteen pounds) during the next year.[24] Of those who ended the year at their goal weight, only half said their weight had been stable throughout the year, suggesting that the other half were weight cyclers, not true weight-loss maintainers.

People in that study were more likely to regain their lost weight if they had lost more weight initially, had maintained the loss for less than two years, had a history of weight cycling, or started the study year intending to lose additional weight. A long history of maintenance increased the odds of continued success, but no group was perfect. Even among people who had maintained their weight loss for at least fifteen years, 20 percent gained more than five pounds over the following year. Research on successful maintainers can tell us how they keep the weight off—typically by counting calories, exercising daily, and staying on a diet for the rest of their lives. That's a tough job, and the people who do it make managing their weight a central priority, at the cost of other goals that could be receiving that time and effort.

Many of us have an acquaintance who has lost weight and kept it off for years, but prescribing that approach as a general rule for everyone is a mistake because most people can't stay on a diet for decades. In addition, some people's bodies hang on to weight so intensely that nothing short of prison-camp conditions could ever make them thin. Besides, special cases like successful dieters tend to be easier to bring to mind. Anyone who mentally sorts through all their acquaintances will probably find vastly more people who've lost weight and then regained it than successful long-term weight losers.

To make claims about typical diet experiences, we need to look at a large number of people instead of a few inspirational stories. One analysis of multiple studies examined the long-term outcomes of weight-loss attempts.[25] A year after the start of a diet, on average, people had lost 6 percent of their starting weight—and were as thin as they were going to get. For example, someone who had gone on a diet at 200 pounds would be down to 188 pounds when the weight started coming back. After that, the average person slowly but steadily regained, on track to fully reverse the weight loss after five and a half years. There was no sign that the rate of weight gain was slowing down at four years after the diet, the latest time that was studied. Another review of multiple studies also found successful weight-loss maintenance to be rare.[26] The authors concluded that only 15 percent of dieters managed to maintain a weight at least 22 pounds below their starting point for three or more years. And the more weight people lose, the more likely they are to gain it back.

Speaking privately, the diet industry agrees. A report for professionals stated, "In 2002, 231 million Europeans attempted some form of diet. Of these only 1% will achieve permanent weight loss."[27] Of course, if helping people lose weight is the business model, a huge pool of repeat customers might be good news for the bottom line.

When research shows that a particular diet "works," we need to check how long the study lasted to know whether the loss is likely to be temporary. Most diet studies last only six months, and few of them follow participants for longer than a year, due to limited funding and perhaps anxiety about documenting the long-term failure rates. Even so, scientific journals publish these short-term studies, and journalists report them, misleading readers who are looking for a permanent solution.

Existing studies overestimate long-term weight-loss success. The most important reason is that people who took part in the initial weight-loss program are often difficult to contact years later. In fourteen long-term studies, researchers located only 33 percent of participants, on average, four or more years after the diet.[28] People who regain more weight are less likely to attend follow-up appointments, presumably because they're ashamed of their poor results, so tracking down a larger fraction of dieters would reduce the reported success rates.[29,30]

Another problem with long-term studies of weight loss is that people who regain weight after dieting but before follow-up often end up going on other diets in the meantime. Imagine a common pattern, in which someone loses weight in 2012, gains it back in 2013, diets it off again in 2014, gains once more in 2015, and diets again in 2016. If researchers studying the long-term outcome of the 2012 diet call in 2016 and find the dieter is below her starting weight, they may put her down as a successful maintainer, when she is really a weight cycler. For all these reasons, one review concluded that better-designed studies are less likely to find that dieting is an effective road to long-term weight loss.[31]

Do We Need Better Diets?

Maybe most dieting is ineffective because we're doing it wrong somehow. Many people believe that only a certain type of diet (low carbohydrate, low fat, intermittent starvation, or whatever) causes people to lose weight permanently. However, clinical studies do not support this belief. In head-to-head comparisons, diets that provide the same number of calories through different types of food result in similar weight loss[32] and regain. Certain individuals may find a particular diet easier to follow than others, but there's no evidence that any of them can permanently lower the brain's defended range.

Maybe if we lost the weight more slowly, we could keep it off. No evidence supports this common belief either, despite its intuitive appeal. One year after weight loss, there was no difference in outcome between people who had lost their weight quickly and those who had lost it slowly.[33]

Some point to snacking as a possible culprit. But there's not much research to back up that belief. In one randomized controlled trial (the gold standard for intervention studies), being required to eat snacks had no effect on people's weight because they ate less at meals to compensate, as a weight thermostat should cause them to do.[34] In addition, observational studies do not show any association of weight gain with eating between meals.

Perhaps the problem is taking a short-term approach to weight loss, instead of focusing on permanent lifestyle changes. Unfortunately, while temporary changes do produce temporary results, the long-term effects of weight-loss programs are disappointing even with persistent support. The Diabetes Prevention Program provided individual coaching for three years, with one coach per twenty dieters. The coach delivered an initial sixteen-session curriculum, followed by group classes and motivational campaigns, along with interventions tailored to remove barriers identified by each person. Supervised exercise sessions were offered twice a week. With all this attention, the participants lost an average of fifteen pounds by the end of the first year and had regained almost all of it five years later, leaving them an average of four pounds lighter.[35]

The Look AHEAD program had similar results.[36] This state-of-the-art behavioral weight-loss trial compared standard lifestyle advice for diabetics with an intensive intervention in more than five thousand overweight or obese patients who already had type 2 diabetes (the kind that's related to obesity) when the study started. The intensive program was well named, providing patients with more than three hundred contacts with dietitians, psychologists, and exercise specialists over eight years. It started with 108 group or individual sessions in the first year, continued with 24 sessions and 12 phone calls per year for years two through four, and ended with 24 sessions per year in years five through eight. In an unusual twist, even the control group lost a bit of weight by the end of the study, perhaps because of the motivation provided by having diabetes or the researchers' decision to prescreen participants for compliance with instructions. With all that effort and expense, after the eighth year, the intensive group had lost an extra 2.6 percent of their body weight, so they weighed 5.7 pounds less, on average, than the controls.[37] About the same number of people gained weight over the study (26.4 percent) as lost at least 10 percent of their weight (26.9 percent). Those are the best long-term results ever reported for a behavioral weight-loss program, and they are underwhelming to say the least. In short, the problem isn't that we're dieting the wrong way. It's that dieting isn't getting us anywhere.

Dieting Can Make Us Fat

The popularity of dieting has soared in recent decades.[38] In the 1950s, 14 percent of women and 7 percent of men in the United States were trying to lose weight. By the mid-2000s, 57 percent of women and 40 percent of men had been on a diet in the past year.[39] That's not simply a result of the increase in obesity, as almost half the female dieters in this survey were within the normal weight range. Even among underweight women, 13.5 percent wanted to lose weight. A survey of more than sixteen thousand university students in twenty-one European countries found that 44 percent of women and 17 percent of men were trying to lose weight, though few of them were overweight.

Among European students with weights in the middle of the normal range, 60 percent of women considered themselves too fat.

But the evidence shows that instead of leading to reliable weight loss, dieting puts many people on the path to weight gain. Within four years, two out of five dieters end up heavier than they were before they lost weight. Deliberate attempts to become thinner strongly predict weight gain over the long term, even when researchers take initial weight, diet, and exercise habits into account.[40] Because dieting is more likely to make people fatter than thinner in the long run, the increased prevalence of dieting over the past few decades may itself be one cause of the increase in the obesity rate.

Take a moment to think about that. Doctors routinely recommend a treatment for obesity (dieting) that is likely to result in their patients' gaining more weight. After one of my talks, a doctor told me that it felt unethical not to recommend dieting to her obese patients, but when I asked her about the ethics of prescribing a treatment with such a high failure rate, she fell silent. Although I have no doubt that doctors want to help their patients, medical professionals should pay attention to the effectiveness of their recommendations instead of indulging in wishful thinking. Like the rest of us, though, they are often misled by the short-term success of dieting and discount the long-term failure rate. Plus they just don't know what else to do.

Doctors and the rest of us could start by worrying less about weight. The intense focus on weight loss suggests that obesity must be uniquely deadly, but that's far from true. Low fitness, smoking, high blood pressure,[41] low income,[42] and loneliness[43] are all better predictors of early death than obesity, when considered individually. Public health efforts that increased physical activity and decreased social isolation would save vastly more lives than battling the brain's weight-regulation system.

The war on obesity has never been a fair fight. As one doctor bluntly put it, "Rather than showing what does work for preventing and treating obesity, research to date shows us clearly what does not."[44] Our current tactics, as a society and as individuals, have caused us to lose ground over the past several decades, and the collateral damage has been extensive. We can't continue doing the same things while expecting different results. Instead we should be asking how we can change strategies to begin moving in the right direction.

I found my own way forward through understanding how psychology and neuroscience explain the common experience of weight gain, loss, and regain. This knowledge made me realize that my brain responds to weight loss in a normal way, so I shouldn't feel like that's my fault. Until I resolved to stop dieting, I hadn't realized how much routine stress I was feeling during every meal, how badly the belief that I needed to be slender was damaging my self-image, or how much mental energy I was wasting on trying to control my eating. All in all, I consider leaving dieting behind to be one of the best decisions I've ever made.

2

Willpower Runs Out

Dennis Asbury struggled with the growing realization that keeping the weight off was harder than he'd expected. At first he was certain he'd mastered his new lifestyle. Every morning he got to work early to spend an hour walking in circles around the unfinished basement, and he ate sixteen or seventeen hundred calories a day. But by the second year, thoughts of food had become a constant distraction. Sometimes he caught himself envying heavier people on the street because they could eat whatever they wanted. Beyond the deprivation, though, he was starting to feel that "All I'm doing is being obsessive about counting the calories. I'm being obsessive about the exercising. I'm working, and I don't have any time to do anything that I might like to do otherwise." To put it another way, his willpower was becoming exhausted.

His weight started creeping up again in the fall of 1999, two and a half years after the diet ended. Work had become increasingly chaotic, with colleagues getting fired in the middle of the night. In April 2001, Dennis hit two hundred pounds and stopped counting calories. The transition felt like "the old joke that mixed emotion is watching your mother-in-law drive over a cliff in your new Cadillac." The relief at being able to eat freely was shadowed by the feeling that he was a failure, someone who'd had thinness within his grasp but couldn't hang on. He ate everything he'd been missing, and then he ate some more, feeling completely out of control. "I guess it was binge eating.

That's the only thing I would know to call it." Within a year, he was back to his starting weight. The situation at work had improved, but he wasn't happy with himself.

When he retired at age fifty-nine in 2010 he decided to use his new freedom to lose the weight for good. Life had gotten much less stressful for Dennis and his wife, Ruby. They bought her childhood home near the Blue Ridge Parkway as a summer place, returning to their North Carolina house in the winter. Again he started counting calories and walking for hours every day. His weight dropped from 266 pounds to 150, and his doctor took him off the drugs that had been controlling his high cholesterol and blood pressure.

As time wore on, though, he found himself dwelling on all the foods he couldn't eat: no more pizza, pasta, or chocolate cake. Managing his calorie budget required constant vigilance. He had to make sure he ate most of his allotment in the evenings, or he'd be too hungry to sleep. That meant starving himself until afternoon and then trying to just nibble a little to take the edge off. He couldn't afford to lose control even for a moment, because if he slipped, he knew he would binge.

The real trouble started about four years after the diet ended, in the summer of 2014. His logbook showed that he had walked four thousand miles, and he was getting tired. This time he struggled repeatedly to get back on track after he'd slipped. Some days he stuck to the plan, while other days he binged. Seven months later, when I spoke with him, he was up to 172 pounds and starting to accept the idea that a steady diet of deprivation wouldn't work for him in the long run, despite the impressive store of willpower that had allowed him to succeed in eating less than his body wanted for a total of nine long years.

Like Dennis, many people think of dieting as a simple conflict between willpower and temptation, but it actually involves multiple brain systems, each with its own priorities. The energy-balance system or "weight thermostat" tries to keep weight within the defended range, while the reward system pushes us to eat calorie-dense foods, and the executive system, which is responsible for deliberate self-control, intervenes in favor of our deliberate intentions, typically to eat less. The reward system determines our enjoyment of food and how hard we (and other animals) are willing to work for it. Without those rewards, we might never get off the couch, but this system is prone to short-term thinking and impulsiveness, as shown by its key role in drug addiction. Repeated activation of the reward system can also install habits, which contribute to the difficulty of changing long-standing behaviors. The executive system helps us plan and accomplish goals like weight loss. Although this system is effective over days to months, it works poorly over years to decades because applying it requires deliberate effort, which can't be sustained all the time. The energy-balance and reward systems, in contrast, are always on duty, influencing behavior for the most part without our conscious awareness. When we try to use willpower to maintain a weight loss, we have chosen the wrong tool for the job.

The Voices in Our Heads

When I was first learning to ride a motorcycle, I once left a driveway without having made a definite decision to turn left or right. The two options were still fighting it out in my mind when I reached the road—and continued straight ahead, directly into a bank on the opposite side. I landed on the ground, more embarrassed than hurt. These sorts of fiascos are possible because the brain, like a lot of poorly run institutions, is governed by committee. People tend to think of the brain as a single organ, but it's more like a collection of systems that constantly compete to control behavior. That's why we find ourselves reaching for that brownie we just decided we wouldn't eat. Through a similar competitive process, dieting often ends up leading to outcomes that no sensible person would intend.

Under most conditions, the energy-balance system is very effective, achieving 99.5 percent accuracy in matching calories in to calories out, much higher accuracy than anyone can achieve with a kitchen scale. But its function is strongly modified by the reward system. For our ancestors, killing a large animal or finding a nest full of honey was a rare and wonderful event, so it made sense to overeat at those moments and store as many calories as possible for the lean times that might come later. Needless to say, this strategy doesn't work as well under conditions of abundance, when tempting food is everywhere, and the famine never arrives.

Rewards have the same purpose in the everyday life of adults as gold stars in preschool classrooms: to motivate us to engage in particular types of behavior. The same brain system is activated by food, sex, or achieving a goal—at work or on the playing field. For internal brain rewards, the incentive structure is determined not by teachers but by evolution. Certain activities are enjoyable because they contribute to the survival of the species. People who failed to take care of such important business didn't manage to become anyone's ancestor, because they died young or failed to reproduce. Scientists distinguish between natural rewards, which are key to survival, and artificial rewards, like heroin, which activate the same brain circuits without providing any evolutionary benefit.

The brain's reward pathways have been studied extensively. When activated, neurons in the ventral tegmental area (VTA) deliver a neurotransmitter called dopamine to targets in the striatum (including the nucleus accumbens, perhaps the best-known reward area), amygdala (a key regulator of emotion), and prefrontal cortex (part of the executive-control system). This activation increases our motivation to seek out rewards and makes them grab our attention. Activity in dopamine neurons helps animals, including people, learn to predict when and where they can find rewards in their environment.

Substantial weight loss increases not only hunger but also the rewards of eating. A hormone called leptin, which is produced by fat cells and travels in the blood to the brain, tells the hypothalamus how much stored energy is available. Leptin also acts directly on the VTA to reduce the reward value of food, making eating less attractive to people when they have adequate fat stores.[1] In contrast, as weight loss reduces fat stores and thus leptin levels, eating becomes more rewarding. When people who've lost weight see a picture of delicious food, the brain shows more activity in reward-related areas than it would if their weight were within their defended range.[2,3] Similarly, calorie-restricted rodents are willing to work harder for food than well-fed ones, and they learn more readily to hang out in the area where food is delivered.[4] Calorie restriction seems to sensitize the brain to all forms of reward, not just food, as drug addicts are more likely to relapse when they're hungry.[5,6] Even so, food provides an intense reward: cocaine-addicted rodents work harder to obtain sugar than cocaine under some circumstances.[7]

When the same behavior repeatedly leads to a reward, that action may become automated as a habit. This process can lead us to eat in mindless ways that are hard to notice or control. We all overestimate how often we're making deliberate choices instead of acting by habit. Because choices are costly, the brain's default is to apply the least possible effort to any situation. A typical person makes two hundred to three hundred food-related decisions per day,[8] so a large fraction of them are bound to be habitual.[9] Habits easily override our conscious intentions, as happens when we end up at a familiar destination like home or work when we intended to go somewhere else.

In mice, eating habits seem to be controlled by a brain area called the lateral hypothalamus, which is interconnected with the reward system. Some neurons in this area project directly to the dopamine neurons of the VTA. When this pathway is stimulated in mice, they continue to eat after they're well fed.[10] They also approach a location where they have been rewarded with sugar in the past, continuing to do so when the sugar is no longer available. The mice are so strongly drawn to this location that they'll approach it even after they've received an electric shock there.

The mice's behavior resembles addiction. When that brain pathway is blocked, mice still eat if they're hungry but won't look for sugar when they're not. Other neurons in the lateral hypothalamus signal when rewards are unexpectedly received or omitted, which may train reward-driven feeding habits. This work suggests that the reward system controls eating behavior both directly, by motivating animals to eat tasty food when they aren't hungry, and indirectly (and perhaps more powerfully), by driving the development of eating habits that continue after they stop being enjoyable or even become unpleasant.

Sound familiar? In our environment—which provides intense motivation for overeating while the culture insists that everyone should remain extremely thin—many people have developed a difficult relationship with food. Every meal, convenience store, and vending machine kicks off a battle against the temptation to eat food that we don't need or perhaps even want. Some people learn to feel guilty whenever they eat. Others make elaborate rules to tell them what types of eating are acceptable, in the hope of gaining control over their unruly bodies. Such uncertainty about how to feed ourselves is a new phenomenon in human history, arising just since our grandparents' generation. Back then, few people experienced eating as a white-knuckled struggle with the enemy.

The Bank of Willpower

When we choose to engage in the struggle between impulse and eating, our main weapon is willpower. Advisers from relatives to doctors suggest that willpower is the key to weight loss, but that approach has serious limitations. Willpower engages the brain's executive system, a set of brain regions that control actions requiring deliberate effort.[11] This brain system is needed whenever we turn off our automatic pilot and do something unfamiliar or difficult. Without it, we would never figure out how to solve a complicated problem or do today what we could put off until tomorrow. As a general rule, new or hard tasks require executive function, while anything that we do often or easily is likely to be a habit.

Parents of teenagers have probably heard of one important area in this system, the prefrontal cortex. The late development of this brain region, which isn't finished until the mid-twenties, explains the poor impulse control of adolescents. While executive function is important for self-control, its role is much broader, including the ability to set goals, plan a sequence of actions and carry them out, and resist distraction along the way. Executive function is important for intelligence, particularly cognitive flexibility and the ability to hold multiple pieces of information in memory while working on a difficult problem. It's also a key component of the ability to focus attention on a particular task. Finally, people with strong executive function can regulate their emotions well and thus tend to have good social skills. All these abilities develop together during childhood, and someone who is good at any one of them is likely to be good at the others too.

Compared with our primate relatives, people have a huge prefrontal cortex, and our executive function is similarly impressive. But as an evolutionarily recent development, it's still not all that good. In contrast to the energy-balance, reward, and habit systems, using executive function takes effort, which means that the system needs frequent breaks to recharge. Executive function is impaired by stress, sleep loss, loneliness, or lack of exercise—all factors that increase the odds of obesity. Because low-calorie dieting leads to the production of stress hormones, being on a diet can directly impair the willpower that is needed to succeed in resisting temptation.[12]

Willpower can also be depleted by overuse. Pushing effortful action too far can use up willpower temporarily, as it is a limited resource in the brain. In a series of studies, people who were asked to do one task requiring effort to finish were less persistent in carrying out a seemingly unrelated task.[13] For example, participants were asked to eat radishes and then complete a puzzle. They didn't know that it was impossible to solve. To make the radish eaters feel especially deprived, the radishes were presented alongside freshly baked chocolate chip cookies that the participants were told not to touch. Radish eaters gave up on the puzzle sooner, in eight minutes on average, less than half as long as the people who were allowed to have cookies. That was true whether or not the participants were dieting. Similarly, people asked to cross off every instance of the letter *e* on a page from a statistics textbook showed less persistence in watching a movie of a wall and a table, in which nothing ever happens. Doing anything that requires effort makes it more difficult to do something else that requires effort until willpower has had a chance to recover.

A lot of evidence indicates that different tasks call upon the same willpower reserves. That is, efforts to resist tasty food leave people with less ability to stick to their household budget and less patience for their child's new drum set. Willpower depletion is familiar to anyone who's ever broken a diet by grabbing an ice cream bar in the convenience store at the end of a long day, without being at all hungry. No matter how willpower got used up, when it runs out, it's gone from all aspects of life. In addition to being inclined to overeat or overspend, people who are short on willpower are less likely to finish a tough project at work and cope poorly with minor stress. Another way to look at the same data is that mental fatigue causes people to become less motivated to complete difficult projects, causing them to voluntarily switch to an easier task. The effect is the same either way: spending too long on effortful tasks leads people to shift their efforts to more immediately rewarding pursuits.

Most people overestimate their willpower and its influence on their daily lives. One meta-analysis (a way of combining multiple studies to reach an overall conclusion about research in a field) on the effects of willpower traits on behavior found evidence for a lot of self-delusion.[14] People's answers to questions about their self-control are much better correlated with how they imagine they will behave in the future than with how they actually behave. In particular, self-control barely influences eating and weight, while it has twice as much influence on school and work performance. The authors conclude that some behaviors are more controllable than others, and that eating is one of the least controllable. A biologist would expect that outcome, because eating is critical to survival in a way that professional or educational achievements are not. That suggests it is probably not regulated by our unreliable systems for deliberate cognitive control. Instead, eating is regulated through multiple feedback pathways that act mostly without our conscious awareness with the goal of stabilizing body weight, especially against starvation.

Contrary to what we might expect, the meta-analysis found that self-control most strongly influences automatic or habitual behavior, with only small effects on tasks requiring deliberate control to complete. That result suggests the most effective use of self-control is to help us form good habits, which can then be performed again and again without much effort. This clever strategy gets around the key problem with willpower, its lack of persistence.

When willpower is depleted, people become more likely to behave according to their habits, whether those habits match their goals (going to the gym) or not (drinking too much).[15] For example, when students were asked to spend two days doing every action that was not essential with their nondominant hand, they increased performance of their strong habits—desirable and undesirable alike—with little effect on their weak habits. For people who have cultivated habits that serve their goals, that's good news. For people whose habits could use a tune-up, perhaps not so much.

Some psychologists believe that willpower is like a muscle, in that its function improves with repeated use. Based on this model, we might expect that exercises to increase willpower in one area might also increase our capacity to carry out other difficult tasks. Doing several unrelated tasks requiring effort in a row might be an even more effective means of improving willpower. People get better at anything they do repeatedly, of course, but the idea is that increasing willpower in one domain also increases its abilities in other domains. This effect is particularly strong in children, as self-control develops more quickly in those who practice it.

Perhaps surprisingly, one of the best ways to improve executive function is through physical exercise. Intervention studies show that physical activity increases prefrontal cortex activity and executive function in the elderly and also protects against cognitive decline and dementia. Physical activity seems to be important for brain function across the lifespan. Increasing fitness improves executive function in young people, and physically fit children get better grades. In one study, sedentary elementary school students were randomly assigned to participate in an exercise program for twenty or forty minutes a day over three months.[16] The exercise group showed improved planning ability and increased frontal cortex activation during a task that required them to inhibit an automatic response, a process that requires executive function. Although we can build up additional executive function over the long term, few of us have so much capacity that we never run out.

Willpower Is a Short-Term Solution

In real life, willpower always fails eventually. Relying on it to control weight means putting our goal at risk every time something goes wrong in our lives. This approach also falls apart if people don't have good choices available. If a neighborhood has no vegetables for sale, and the streets are dangerous, all the willpower in the world is not going to keep most people thin. The idea is especially ridiculous for children, who have not yet developed their full executive-control abilities. Is it really a three-year-old's responsibility to resist the pull of four thousand food ads per year and make smarter choices? Changing the environment would provide much more effective solutions, for instance by banning food advertising aimed at children. All the same, misguided obesity-prevention programs have tried to teach children as young as three to use willpower to balance calories in against energy out. Needless to say, such efforts are doomed to fail—and to do real harm in many cases.

In the struggle of the energy-balance system versus the reward and habit systems for control of eating, the modern environment has tilted the balance toward the reward system by making it so easy and cheap to get calorie-dense food. Dieters try to use willpower to regain control, but the executive system functions best when we don't ask it to work long hours. As a result of these brain quirks, for most of us, dieting ends up as a vicious cycle. First we fight against the energy-balance system to lose weight. For a while, we succeed in suppressing it by using willpower to ignore hunger, but eventually we shift our attention to some other problem. As soon as that happens, the energy-balance system, which has been energized by our weight loss, pushes hard for us to eat more and exercise less—not to mention suppressing our metabolism, which willpower can't influence at all—leading to weight gain.

Once we're back within our defended range, the energy-balance system settles back down, returning our hunger to normal. But as a consequence of the time we've spent suppressing and ignoring hunger, we've become less responsive to the signals that tell us to stop eating, which creates an opportunity for the reward and habit systems to step in and take charge, often leading to further weight gain. After a while, for reasons that researchers don't fully understand, the energy-balance system comes to see that higher weight as normal and starts to defend it. Then we decide that we're too fat and start the cycle over again. In this way, an outcome that no one would choose—putting in a lot of effort but ending up heavier than when we started—has become a common experience.

A better alternative is mindful eating, which uses willpower as a temporary intervention to put the energy-balance system back in control. By paying attention to signals of hunger and fullness, we can learn to recognize them more easily. Through this deliberate practice, we develop habits of eating only as much as our energy-balance system commands, an improvement that lasts even when our willpower has moved on to other problems. For this approach to be effective, though, we have to make peace with the weight that the energy-balance system is trying to maintain, which may involve losing less weight than we wanted or none at all. Mindful eating can break the diet-overeating cycle that has trapped many of us, but it won't keep weight below the defended range. Although that's a big hurdle for some people to get over, the payoffs are tremendous. We can learn to eat with relaxation and pleasure, while avoiding weight gain and reclaiming our willpower for more important projects.

People who try to use willpower to keep their weight below its defended range find that their energy-balance system continues to try to push their weight back toward that range for years, every minute of every day. Whenever their willpower is depleted—that is, when they get overworked, stressed, tired, upset, or otherwise in less than perfect health—the energy-balance system takes advantage of that weakness to push them to gain the weight back. The only way to succeed at willpower-driven weight loss is to remain on a strict calorie-restricted diet for life, in sickness and in health, for richer and for poorer, until death do you part. A small number of "willpower athletes" manage to accomplish this feat, but it's a tough way to live, and most people fall off the wagon eventually.

Even success at that daunting proposition comes at a substantial opportunity cost. Because willpower is limited, using it to maintain body weight leaves less of it to spend on other areas of life. From my perspective, most of us could find better uses for our willpower—to improve our relationships with friends, partners, and children, our work achievements, and our contributions to society—rather than spending it on trying to fit into smaller pants. Anyone who says that weight loss should be maintained with willpower is saying that keeping weight down should be everyone's most important priority forever. After careful consideration, I decided that choice doesn't align with my values. Eating well and exercising should take care of maintaining my health. Beyond that, I'd rather spend my time and energy doing something useful for other people than making myself miserable by worrying over my appearance, a problem that our diet-obsessed culture encourages.

3

How Diets Lead to Weight Gain

In 1945, a group of conscientious objectors from the U.S. military volunteered to lose 25 percent of their body weight over six months and then regain it at various rates.[1] The experiment was intended to help researchers determine the best way to rehabilitate starving people after the war. The thirty-six men were placed on a diet that provided 1,570 calories a day, about half of what they'd been eating before, and were required to walk twenty-two miles a week.

Even on that restrictive diet, two of them stopped losing weight before they reached the target. They were suspected of cheating and their meals were reduced still more, to fewer than one thousand calories per day. To cope with their hunger, they chewed up to forty packs of gum a day. Months later, both were dropped from the study, still insisting that they hadn't cheated. The vast individual variability in weight gain and loss makes that claim plausible.

The men had been chosen from more than two hundred applicants because they were physically and psychologically healthy when the study started, but few of them stayed that way. All the volunteers became ravenous and exhausted due to the physiological effects of severe weight loss. Their strength decreased by 21 percent, their average heart rate dropped from fifty-five to thirty-five beats per minute, and their metabolism declined by 40 percent. With a low metabolism and little body fat, they felt cold all the time. Within a month, they had lost all interest in sex. The first subject quit in a few weeks, after dreaming of cannibalism and threatening to kill himself or the chief researcher, Ancel Keys. After a few days of regular meals at the psychiatric hospital, he seemed normal again, so the doctors released him from the hospital and from the study.

Over time, the volunteers became as obsessed with food as anyone with an eating disorder. On the day the war ended in Europe, one man failed to note the victory in his diary, instead complaining of the crushing psychological consequences of having to wait in line too long for dinner. Another man spent hours staring at the photos in his cookbook collection and fantasizing about food. As the experiment wore on, the men became irritable and sometimes fought over scraps of food. They began to perceive their own bodies as normal and everyone else's as too fat, a mind-set typical of anorexia patients.

When the volunteers had lost as much weight as the study dictated, their portions were increased, but they still wanted to eat a lot more than they were given. In the third week of recovery, in despair as his hunger remained acute despite the extra calories, one volunteer cut off three of his own fingers with an axe. Because the researchers needed his data, they granted his pleas to remain in the study. At first, he convinced himself that the axe had slipped by accident, but later in life he conceded that the blow was probably an attempt to end his participation with honor, as he couldn't face becoming a quitter. Supporting that interpretation, a week earlier he'd managed to crush a finger when a car jack fell on him.

Once the experiment ended, the men ate more than five thousand calories a day, continuing to shovel in food even when their stomachs were stuffed. For decades afterward, they suffered from binge eating, depression, and mood swings. They harbored a persistent anxiety that their food might be taken away again someday. At the last follow-up, in 1998, researchers located nineteen of the men. They reported that after the experiment, they had gained an average of twenty-seven pounds above their original weight, though only three men stayed at those higher weights for the rest of their lives.

Another group of researchers studied Canadian veterans of World War II, comparing those who were captured and then starved against their fellow veterans with equivalent combat experience.[2] About a half century after the war, 45 percent of the former prisoners said they were moderate or heavy binge eaters, compared with 17 percent of the other veterans. The more weight the prisoners had lost during their captivity, the more likely they were to report binge eating later on. The experiences of all these men suggest that starvation can change eating behavior and attitudes toward food for a long time afterward.

Your Brain Thinks Diets Are Starvation

Many dieters find themselves behaving like starvation victims. Indeed, the protocol in the 1945 study—1,500 calories and an hour of exercise a day—is what most of us would call a sensible weight-loss diet. That's what I did for most of my adult life whenever I thought I was too heavy. Dennis Asbury required that much food and exercise just to maintain his weight within the normal range.

We may believe that losing weight is a good idea, but our brains often don't agree. Someone who starts at four hundred pounds and then loses one hundred pounds has lost 25 percent of his body weight, just like the volunteers in the starvation study did, and his brain changes his physiology in much the same way. Most dieters who've lost a lot of weight respond to food restriction as if they were starving to death, by developing an intense drive to eat, as Dennis discovered.

Scientists have known about this psychological response for almost half a century, since Jules Hirsch reported the results of his early weight-loss studies in 1968.[3] Rockefeller University had an inpatient facility that allowed scientists to observe what happened when obese people lost weight. The participants were highly motivated, as they were willing to spend sixteen to twenty weeks living away from home on a liquid diet that provided only six hundred calories per day. Hirsch and his colleagues had long been puzzled by the fact that the people who lost weight in their program always seemed to regain it when they went home, despite claiming that they wanted desperately to remain thin. To figure out what was going on, he put together a multidisciplinary team that could evaluate everything from the participants' physiology to their mental state.

The psychiatrists found no behavioral or psychological problems in the obese participants when the study began. To no one's surprise, the dieters were grumpy and hungry during weight loss. Other psychological symptoms, though, persisted after weight loss ended. The formerly obese people were mentally and physically sluggish. They showed increased anxiety and depression, along with an intense obsession with food, including dreams and waking fantasies about food and eating. Perhaps as a consequence, they also admitted to breaking their new diets repeatedly, often eating in near-uncontrollable binges, as Dennis Asbury found himself doing. The researchers were struck by how much these responses resembled the psychological and physiological consequences of starvation.

Dieters Get Fatter in the Long Run

Like starvation, dieting is a strong predictor of future weight gain. Fifteen long-term studies, which followed dieters for one to fifteen years, all found that dieters are more likely than nondieters to become obese.[4] That's true in men and women, across various ethnic groups, from childhood through middle age. The effects are strongest in people who started dieting when they were in the normal weight range, perhaps because weight loss reduces their muscle mass the most. When dieters regain their lost weight, they gain fat faster than they gain muscle, which leads them to replace muscle with fat or to recover their strength at the cost of putting on extra weight. People and laboratory animals with a genetic tendency to gain back more weight than they lost may produce additional fat cells during the regain process, providing a place to store more weight.[5] These results suggest that dieting may be counterproductive for many people, as it leads to weight gain, on average, rather than loss. At best, for people (like me) who don't regain more weight than they've lost, it's merely useless.

Because people in such studies aren't randomly assigned to diet, some experts argue that the relationship goes in the other direction: people who are genetically prone to weight gain are more likely to try to lose weight.[6] That line of reasoning presumes that normal-weight people who start dieting were already gaining weight beforehand, due to their susceptibility to an environment of food abundance. No doubt that's true in some cases, but it strikes me as unlikely to be the whole explanation. Most of the teenage girls I knew didn't start dieting because they were gaining weight. Instead we wanted smaller bodies than we naturally possessed.

To sort out the influence of genetic predispositions from dieting itself, researchers did a nine-year study of more than four thousand twins, from the ages of sixteen to twenty-five.[7] Among identical twins who started at normal weight, when only one twin was a dieter, that twin was more likely to gain weight than the nondieting twin over the next nine years, indicating that dieting can increase weight gain after accounting for genetic background. The difference in weight gain between dieters and nondieters was even larger among fraternal pairs, though, supporting the idea that dieters may also have a higher genetic tendency toward weight gain. A single episode of deliberate weight loss in this study increased the odds of becoming overweight by a factor of two in men and a factor of three in women. Women with two or more episodes of dieting were five times as likely to become overweight by age twenty-five.

Another way to test this idea is to look at people who had an external motivation to lose weight, like the starvation-experiment participants. In his early twenties, Michael Vassar went on a diet, although he wasn't worried about his weight. Instead, as a futurist working at the Singularity Institute for Artificial Intelligence in San Francisco, he was intrigued by research showing that long-term calorie restriction can extend lifespan in insects and rodents. Along with other enthusiasts, he hoped to show that this approach would work in people as well. To do so, he aimed to eat 1,400 calories per day, reducing his weight from 150 pounds to somewhere in the 120s.

He developed a series of rituals to control his eating. To make sure that he didn't eat thoughtlessly, he required himself to state his intent to eat out loud before he took the first bite of anything. If he was still hungry, he had to wait half an hour before eating again. With experience, he became an expert on hunger, aware of the trade-offs of one food choice over another. He learned to avoid muffins and pastries, which were too difficult to eat in moderation. When he ate less, he discovered that food tasted better. Living in the Bay Area made the experiment easier, as "life-hacking" projects like trying to manage on less sleep were routine in his social circles.

The diet definitely had drawbacks, though. Because his metabolism was suppressed, he felt cold almost all the time, except when he was directly in the sun on a warm day. His body temperature was ninety-five to ninety-six degrees, and his sex drive disappeared. Eighteen months after beginning the diet, he moved to Kazakhstan, where winter temperatures average four degrees below zero Fahrenheit, to work with the Peace Corps. He decided that such a cold climate was incompatible with calorie restriction, so he stopped dieting and allowed himself to eat normally. He regained all the weight that he'd lost and more. Discussing his experiences fifteen years later, he says his current defended range centers around 165 pounds, though he weighs 175 pounds at the moment. Losing about 20 percent of his body weight appears to have led Vassar's brain to defend a higher weight range, even though he had no history of concern about his weight.

A history of repeated dieting starting from a normal weight is also typical of people who participate in weight-conscious sports like boxing or wrestling. Such dieting presumably has no connection to a genetic predisposition toward obesity. Elite athletes who competed for Finland in those sports gained more weight by age sixty than their peers who had competed in sports that didn't require weight regulation.[8] The dieters were three times more likely to be obese.

How Diets Promote Weight Gain

Why would dieting make some people gain weight? The explanations relate to a central truth about diets: when people lose a lot of weight, from any starting weight, their brains declare a starvation state of emergency. The solution to that emergency is to gain back the weight—and perhaps a little extra reserve to get them through the next famine.

One reason that dieting might lead to weight gain is that dieting is stressful. From the brain's perspective, dieting is starvation. As we'd expect in that case, calorie restriction produces stress hormones, in people[9] and also in mice. Across a wide variety of species, stress leads to weight gain, and particularly to increases in abdominal fat, which is the type of fatness most strongly associated with medical problems. Yet stress is also a consequence of weight gain, forming a vicious cycle in which people gain weight, making them (and others) unhappy with their bodies, and the resulting stress causes them to gain yet more weight.

Popular opinion has it that people gain weight because they lack self-control, but little evidence supports this idea. The research suggests instead that attempts to exercise strict control over food intake tend to backfire, often leading to weight gain by strengthening the drive to eat. (No one ever said that biology was fair to dieters or anyone else.) If we think of increased hunger as a response to involuntary starvation due to a famine, that motivation to eat makes more sense.

In a large body of work over half a century, psychologists Janet Polivy and Peter Herman studied two types of eaters. Controlled eaters routinely watch their weight. They count calories, keep food diaries, and think carefully about the consequences before choosing what to eat. They often feel guilty when they eat something that wasn't in their plan. Sounds great, right? That's exactly what most weight-control advice tells us we should do, whether it comes from annoying relatives or newspaper columnists or the family doctor.

Unfortunately this advice doesn't work for a lot of people. The constant preoccupation with food uses up willpower and reduces performance on other tasks requiring executive control, like doing a good job at work or parenting.[10,11] Controlled eating also reduces the influence of the brain's energy-balance system by teaching people to rely on external cues to regulate their eating. In our modern environment, the vast majority of those cues say "Eat! Eat! Eat!" After all, marketers designed a lot of them to get customers to buy more food. When we exhaust our willpower and ignore our hunger, the reward and habit systems take control of eating choices.

In contrast, intuitive eaters are sensitive to their bodies' signals, so they eat when they're hungry and stop when they're full. The research papers refer to controlled and intuitive eaters as "restrained" and "unrestrained" eaters, respectively. I don't use those terms because an unrestrained eater sounds like someone out of control who spends his days gulping down Doritos and cartons of ice cream. That's a completely misleading impression. Instead, because of their internal focus, intuitive eaters are less vulnerable than controlled eaters to advertising, super-sizing, stress-induced eating, and the all-you-can-eat buffet.[12] They are less likely to become overweight,[13,14] maintain more stable weights over time, and spend less time thinking about food.[15] Long-term dieters, in contrast, are more likely to eat for emotional reasons or simply because food is available, which makes eating less responsive to the body's calorie needs.

Your Inner Rat

Another reason diets lead to weight gain is that starvation makes people more susceptible to binge eating, as Dennis Asbury discovered. From an evolutionary perspective, this makes sense. If food was hard to find yesterday and may be hard to find tomorrow, then we should eat as much as possible today. Although we often attribute human binge eating to complex psychological motivations, rats share the same basic biology.

In a typical human experiment, some volunteers drink a milkshake, while others get nothing.[16] Then both groups taste and evaluate snacks like ice cream, nuts, and cookies. The researchers don't let on that they're interested in how much of the tasty food participants eat, not how much they enjoy the snacks. Intuitive eaters tend to eat fewer snacks after the milkshake, as we'd expect that the brain's energy-balance system would lead them to do. Controlled eaters, on the other hand, eat less when they aren't given the milkshake—and stuff themselves full of snacks after the milkshake, whether the serving is small or large.

This surprising behavior happens because controlled eaters figure they've already blown their diets by drinking the milkshake, so they might as well eat the foods they've been denying themselves. Follow-up studies provide evidence for this interpretation. If the experiment instead begins with a salad, which has the same number of calories as the milkshake but doesn't make controlled eaters feel like diet cheats, then they maintain control in the snacking phase. Drinking alcohol, doing an earlier task requiring willpower, or being emotionally upset also makes controlled eaters lose control.

The resulting alternation between food restriction and overeating may explain why controlled eaters generally weigh more than intuitive eaters. To be sure, though, we must consider the possibility of reverse causality. People may become controlled eaters because they recognize their own tendency to overeat, which could explain why they continue to gain weight because (say it with me now) diets don't work. As genetics influence weight and eating behavior, the most reliable evidence that food restriction causes overeating comes from a study of almost 1,600 identical and fraternal twins.[17] The results are consistent with research that doesn't take genetics into account. Controlled eaters weighed more at the beginning of the study, showed more weight variability, and had gained more weight four years later than intuitive eaters. This association between controlled eating and weight remained significant in twin pairs who shared both genes and childhood environments, with the weight gain difference between a controlled-eating and an intuitive-eating identical twin being three to six times higher than the average difference between twins. Another twin study found no genetic influence on weight variability or yo-yo dieting, suggesting that these traits result from life experience.[18]

In both adolescence and adulthood, controlled eaters are more likely than intuitive eaters to develop binge eating in studies that follow participants over time. This effect is strongest in people with a high genetic risk of binge-eating disorders, as shown in a study of almost 1,700 twins.[19] Teenagers who use the weight-control strategy of fasting on certain days have a higher risk of developing a binge-eating disorder within the next five years than their peers who are controlled eaters but do not fast.[20] Taken together, all this evidence suggests that repeated dieting probably won't lead to permanent weight loss but does increase the odds of future weight gain due to overeating.

The "diet-breaking sinner" effect is familiar from my own history. If I went over my calorie budget for the day, I would usually spend the rest of the day eating whatever I wanted. Dennis Asbury did the same thing. As he put it, "One Snickers bar, and I might as well just weigh three hundred pounds. My mind tends to that black-and-white thinking: either I'm good or I'm not. If I break down and go over my calories, then I can go eat another thousand calories from the pantry." His binges felt out of his control. "I would find myself eating and eating and saying out loud to myself, 'What do you think you're doing? You know you don't want the result of what you're doing,' but I would just keep on going. I don't know who that is in my head who does that to me."

I replied, "It's your inner rat," because binge eating appears to be a common mammalian response to starvation. Scientists can cause normal-weight rats to binge by alternating food deprivation with access to tasty food, like Count Chocula cereal.[21] Under these conditions (similar to intermittent-starvation diets like the 5:2 program, in which participants eat very little two days per week and don't worry about calories on the other five), animals eat more of the tasty food and become willing to work harder for it than animals that have it available all the time. For example, rodents develop severe binge eating after repeated cycles of five days of food restriction, then two days of free access to Oreo cookies. Four days later, after regaining their lost weight, they're given a brief stressor, such as foot shock or being able to see and smell Nutella but not eat it for fifteen minutes. After this treatment, these animals eat twice as much of the Oreos as animals who received the stressor alone. Just a small taste of Oreos can trigger deprived animals to binge on their regular chow if nothing else is available.[22]

Researchers have identified changes in the brain's dopamine system and other neurotransmitters, which probably change the way that these animals respond to rewards. As in people, the outcomes of these treatments among rats are individually variable, with some animals completely unaffected. Both genetics and early life experiences seem to underlie these differences in vulnerability.

For an animal in the wild without enough food to maintain its body weight, the top priority is getting more. The resulting brain changes reflect that need, increasing the intensity of the animal's motivation to seek out and eat food. In another set of experiments, adult rats were food restricted until they lost 20 percent of their body weight and then maintained at that weight, which required them to eat less than 70 percent of their normal calorie allotment.[23] These food-restricted rats showed stronger reward responses, not only to food but also to various drugs of abuse. They learned where to find drugs more easily and were more persistent in seeking drugs after they were no longer available. Food restriction also seemed to make sugar water act more like a drug of abuse in their brains.

This research suggests that some food experiences may lead to changes in the brain that make overeating easier, and that repeated dieting might make such changes more likely. That conclusion does not mean, however, that most obese people are food addicts or that addiction to junk food is the main cause of the recent rise in obesity.

Most obese people do not show addiction-like brain responses to food[24] and are not binge eaters.

Binge-eating disorder, which occurs in people of all weights, is probably the closest parallel to addictive behavior around food. A key characteristic of addiction is that addicts continue to want something that they no longer like. Binge eaters continue to eat when they're not enjoying it, despite negative consequences and often triggered by stress, similar to the way addicts take drugs. People with binge-eating disorder are susceptible to other addictions[25] and are more likely to have relatives with drug addiction, suggesting a shared genetic vulnerability.[26] Successful treatment of binge eating may lead to weight loss if the disorder has maintained the person's weight above its defended range. Other recovered binge eaters do not lose weight because their defended range has increased as a consequence of repeated overeating.

When I gave up dieting, I wanted to remove the constant drain on my willpower and stop wasting effort on failed attempts at weight control. To my surprise, I ended up recovering much more than some free time. Once I gave myself unconditional permission to eat whenever I was hungry, I noticed a major shift in my attitude toward food. Gradually my anxiety about restrictions faded away, and I became more relaxed about leaving food on the plate when I wasn't hungry, even if it tasted good. In my dieting years, I felt compelled to overeat whenever I was "allowed" to, a compulsion that turned out to be driven by my anticipation of the self-imposed famine to come. When that feeling went away, about six months after I gave up dieting, it was like a stranger had taken over my brain. I no longer feel the need to stock up for the coming famine because I know that I will always allow myself to eat if I'm hungry.

Throughout the first year after I stopped dieting, I worked on noticing when I wanted to eat—and also when I didn't. In the early stages, I watched myself continuing to eat when I was already uncomfortable, especially in restaurants and at parties, and neglecting to eat when I was ravenous because that felt virtuous, when I could manage it. Over months, I learned to sense and respond to my body's hunger and fullness cues. My weight stabilized, and I now enjoy food more, while worrying about it less. With practice, controlled eaters can become mindful eaters, as my experience shows. The transition requires time and commitment, but the potential reward is enormous for people whose food has always come with a side order of guilt or shame.

4

The Weight of Beauty

As the twentieth century neared its end, Fiji was a country in transition. The onset of a cash economy brought television to the islands, with unintended consequences for the local culture, including the way people viewed their bodies. Typical of a society threatened by malnutrition, caretakers had traditionally been vigilant for signs of weight loss in children. Any decline in appetite was met with great efforts to encourage the child to eat more. Before the 1970s, eating disorders were rare or nonexistent. Large bodies were considered most attractive, and becoming too thin was a sign of illness, requiring community intervention. Guests were exhorted to "eat, so you will become fat." Diets were unknown, and the idea of losing weight on purpose would have struck most people as ridiculous.

Then in 1995, television programs brought modern ideas that were rapidly adopted by the young. A few weeks after the introduction of television and again three years later, researchers interviewed seventeen-year-old girls enrolled in two local schools in Nadroga, a province known for its gorgeous beaches and rugby-crazed locals.[1] Girls in the two study years had the same average body weight, but their behaviors and attitudes toward food were different. The proportion of girls who reported disordered eating (such as starvation diets or purging) more than doubled, from 13 percent in 1995 to 29 percent in 1998. Although no one in the earlier group had tried vomiting to control their weight, 11 percent of the later group said they'd done it. Those who had vomited were no heavier than those who hadn't, so body image rather than weight drove their actions. In the 1998 group, dieting was common: 62 percent of the girls had tried to lose weight in the past month. So was body dissatisfaction, with 74 percent saying they felt "too big or fat." The thin ideal had come to Fiji, and teenage girls were no longer comfortable in their own skin.

The United States went through a similar transition starting in the 1970s.[2] Earlier, anorexia and other eating disorders were rare, and men and women were equally likely to dislike their bodies. Thin was not yet in. Advertisements through much of the century, published as late as 1984, offered to help women gain weight and become curvier. As food availability increased over the last half century, models, actresses, *Playboy* centerfolds, Miss America winners, and even female cartoon characters became thinner and thinner,[3] while male action figures became ever more muscular.[4] A woman would need a rib removed to achieve the waist-to-hip ratio of the current Barbie doll, and a man built like G.I. Joe would be stronger than professional body builders. The combination of a culture that values these unrealistic bodies and an environment that makes weight gain ever easier has led to considerable conflict between ideals and reality for most people.

Over the same period, women have become progressively less happy with their bodies, leading to growing differences in body image between men and women.[5] Since the 1990s, the sex difference has been large, particularly in teenagers. This change is due to the large number of girls and women who are very dissatisfied with their bodies.[6] Both experimental and correlational studies link excessively thin role models to increases in body dissatisfaction and disordered eating in women.[7] For example, after viewing thin models in laboratory studies, women are less happy with their bodies and increase their likelihood of binge eating, purging, and crash dieting. Lately men have experienced more body hatred and eating disorders, but they haven't caught up to women yet.

Fat Shaming 101

As a culture, we have learned to judge ourselves by the wrong scale. Intense body hatred, which used to be the province of early teenage girls, is now common from elementary schools to nursing homes. Over 90 percent of adult women in the United States are somewhat unhappy with their shape, and about half can't find one positive thing to say about their physical selves. Forty percent of third-grade girls would like to be thinner, and 60 percent of sixth graders have tried to lose weight. The high rate of body dissatisfaction is no surprise: a study of 2,300 girls found that by age ten, 58 percent had heard a friend or family member say they were too fat.[8] Thinness is no protection, as half of underweight teens report hearing critical comments about their bodies.

In case we don't manage to hate ourselves enough, our culture is ready to pick up the slack. Weight-based prejudice is notable for its openness. People often feel comfortable making direct comments about other people's bodies, ranging from unrequested advice on weight loss to mooing at pedestrians out of car windows. Many of us are unembarrassed to criticize someone's weight to a group of friends, though we might avoid making racist remarks in such company. For example, bioethicist Daniel Callahan wrote an editorial advocating the shaming of obese people to discourage weight gain, despite admitting, "That will not much help most of those who are already overweight or obese. But beyond marginal improvements, most of them are already lost."[9] We might expect a professional to notice the ethical problems raised by mistreating more than half of the U.S. population for the supposed good of the rest.

Children as young as three repeat popular stereotypes suggesting that the obese are lazy, stupid, unsuccessful, and lacking in self-control.[10] In adult life, many studies document widespread bias in actions and beliefs about obesity, with important consequences in areas like education, employment, and health care. For example, women who are sixty-four pounds above average weight are paid 9 percent less than women in the normal range, equivalent to three years of work experience. Wage discrimination against the obese occurs across a variety of countries, ethnic groups, and professions. In laboratory studies, candidates with identical qualifications are less likely to be hired or promoted and are offered lower salaries if they are described as obese. Perhaps as a consequence, obese people are less likely to be employed. Obesity thus increases the risk of poverty, which in turn increases the odds of future weight gain.

Weight bias has increased recently. In the mid-1990s, 7 percent of adults had experienced weight-related discrimination; a decade later that number had increased to 12 percent. Among people who are moderately to severely obese, 40 percent experienced weight discrimination, such as being denied jobs, housing, or medical care based on their appearance.[11] Weight bias from teachers begins in elementary school and continues through graduate school. Obese students are less likely to be admitted to college than their thinner peers with the same qualifications. Such discrimination is legal in most parts of the United States, as weight is not a protected category under the law.

Medical professionals have strong weight-related prejudices, with 69 percent of people in the overweight or obese categories reporting discrimination from doctors.[12] Only family members were more likely to be a source of weight-related stigma. For women, doctors' prejudice begins when they are as little as thirteen pounds outside of the normal range, while men start to experience medical discrimination at around seventy-five pounds above the normal range. The biased beliefs among medical professionals in many countries—that obese patients are uncooperative, hostile, and dishonest—lead doctors to prescribe weight loss for conditions that have no relationship to body fat. As a consequence, obese people are reluctant to go to the doctor.[13] When they do, they get lower-quality care, receiving fewer preventative screenings and less treatment overall. Because early diagnosis and prompt treatment influence health, receiving poor medical care may be part of the reason that obese people are more likely to get sick than their lower-weight peers. Indeed, people who report experiencing weight discrimination have 31 percent higher risk of death than their peers, after their current health, weight, exercise habits, smoking, and depression are taken into account.[14]

We can see why doctors and nurses might be tempted to blame their patients for failing to lose weight. The alternative is to admit that advising obese people to spend the rest of their lives on a diet might do more harm than good. Doctors (and the rest of us) tend to blame people for their medical conditions when they are perceived to be responsible, for example, in drug addiction. That perception is unfair for obesity because individuals don't control many factors that contribute to weight gain, from their genetic heritage to the body's starvation response to the availability of fresh produce or safe places to walk. Why would anyone choose to gain weight in a society that treats obese people so badly?

A Vicious Cycle

Such widespread misery would be a concern all by itself, of course, but this type of blame is especially poignant because body dissatisfaction is a strong predictor of weight gain across the lifespan. In one study of more than six thousand adults (average age sixty-seven), among people who were not obese initially, those who experienced weight discrimination were two and a half times more likely to become obese four years later than those who did not feel that they were treated poorly because of their weight.[15] Among people who were already obese when the study began, those who experienced weight discrimination were more than three times more likely to remain obese. These effects are unlikely to be simply due to stress, as experiencing racial or sexual discrimination had no effect on weight, and they held up when the researchers controlled for starting weights. Another study of almost three thousand British adults over the age of fifty found similar results.[16] People who reported weight discrimination were almost seven times more likely to become obese during a four-year follow-up than those who did not.

This relationship between body dissatisfaction and weight gain has a simple explanation. People who feel bad about their weight are more likely to binge eat, to eat for comfort, and to avoid exercise because they feel physically incompetent.[17] For example, in one study, female college students were asked to read a fake *New York Times* article titled "Lose Weight or Lose Your Job" or a similar article titled "Quit Smoking or Lose Your Job" and then summarize its argument in a five-minute video.[18] Afterward, they sat in a break room with candy and Goldfish crackers. Among participants who read the weight article, those who felt overweight ate more calories than those who didn't, regardless of their actual weights. After reading the smoking article, in contrast, both groups ate the same amount. Thus it appears that thinking about weight discrimination can trigger overeating in women with body dissatisfaction, which long-term studies suggest often translates to weight gain.

Perceived discrimination is linked to psychological distress, especially mood disorders and substance abuse, among people of any weight.[19] Body dissatisfaction is strongly linked to low self-esteem and depression, as well as eating disorders. The most vulnerable to weight stigma are people who have absorbed the societal beliefs that thinness is attractive and fat is shameful. When those people are exposed to antifat messages, they become much less likely to exercise.[20] In long-term studies, higher starting weights predict a reduced likelihood of exercising later, while starting exercise habits have no effect on later weights.[21] Children who are teased about their weight also have negative attitudes about sports, are less active, and spend more time sitting than their peers. In general, obese people who have strongly internalized weight stigma are likely to avoid exercise, socializing, and other activities that could improve their quality of life.

This research shows that body shame promotes weight gain. Yet even public health officials fall into the trap of fighting obesity by stigmatizing obese people. Shaming people to encourage them to lose weight is counterproductive. Few people are motivated to take good care of anything that they hate, and that includes their bodies. Instead, body dissatisfaction leads people to disconnect from their bodies, concentrating on their intellectual abilities and trying to ignore the messy and unsatisfying realities of biology. Distancing ourselves from the body and its signals makes it easier to overeat and more challenging to eat mindfully.

Kelly Coffey knows about the importance of body dissatisfaction for weight gain and loss. She attended her first Weight Watchers meeting at the age of five. After more than two decades of obesity, she had weight-loss surgery in March 2003. The operation presumably lowered her defended weight range, but it didn't make a dent in her binge eating, usually triggered by cookies, doughnuts, or anything containing sugar. She lost a lot of weight and then gained much of it back. As she worked at identifying her triggers and learning to ride out the urges they produced, she started to lose weight again. In 2007, she went into business as a personal trainer, which keeps her focus on sustaining her new habits. She has now maintained a loss of more than one hundred pounds for more than a decade.

We might imagine someone with that history would be drill-sergeant tough and have no time for excuses. Instead, she wrote a charming piece called "5 Things I Miss About Weighing More Than 300 Pounds" (having her own padding for hard chairs, effortless strength from carrying her own weight around all day), which went viral in April 2014. She needs to love her fat body to maintain her healthy behaviors, she says, "because if I live in fear of going back to that place, when I do something counterproductive, I will feel tremendous shame, tremendous guilt, and tremendous fear. And all of those kinds of emotions are in direct opposition to doing things in a healthy way, to taking loving care of myself." Before she learned to value herself, setbacks threw her into a spiral of shame, leading her to overeat and stop exercising, which caused more shame. Now she concerns herself with getting enough sleep, making it to the gym every day, and eating as much as she wants of anything that isn't one of her trigger foods. If she does those things, she says, her body feels good. If she doesn't, no matter what she weighs, she's not happy.

Teenagers and Dieting: A Troubling Mix

Though it may last a lifetime, the tangle of shame, dieting, and binges often begins in adolescence, as it did for me. Soon after I hit puberty, my mother started worrying about my weight. Even though I was right in the middle of the normal weight range for my height, I was looking less thin, so she suggested I'd better start watching my diet. At thirteen, I used the money from my paper route to buy my own low-calorie food, like cucumbers and cantaloupes, which I kept in the basement refrigerator so I wouldn't be tempted by the tasty food upstairs. I'm not sure how I set my goal of 800 calories a day, perhaps from a magazine or other girls at school. I certainly didn't know that amount of food was less than half the energy my body needed—or that my brain would fight back against starvation, whether voluntary or not. When I gained the weight back, of course I blamed my lack of self-control.

As I got older, my approach to dieting became less extreme, gradually moving to 1,000 calories, then 1,200, and eventually 1,500 per day. Although I lost the weight more slowly, I gained it back at the same rate every time. Throughout my teens and twenties, I lost and regained the same ten or fifteen pounds every year.

Repeated dieting is a risk factor for eating disorders, and eventually it caught up with me. When I was in graduate school, in a rocky marriage, and under a lot of stress at work and at home, I started binge eating. I would finish off a half gallon of ice cream or a whole box of saltines with butter, usually at three o'clock in the morning when I couldn't sleep. The emotional pressure to keep eating was intense, even after I had made myself sick. Looking back on it now, I can't figure out why that behavior made me feel like such a terrible person. But for a long time, I was convinced that binge eating was my most shameful secret.

Such problems are widespread, in part because teenagers have a tendency to go overboard with their enthusiasms. When that enthusiasm is weight loss, trouble is often close behind. More than half of teenage girls and a quarter of boys have been on a diet, which tends to predict continued dieting as an adult. Unhealthy dieting behaviors, like skipping meals, occur in 57 percent of girls and 33 percent of boys, while 12 percent of girls and 5 percent of boys have resorted to vomiting, laxatives, or diet pills to try to lose weight. Binges occur in 17 percent of girls and 8 percent of boys.

Pressure to be thin increases the likelihood of both eating disorders and weight gain in teenagers. Dianne Neumark-Sztainer heads Project EAT (Eating Among Teens), a research program at the University of Minnesota to identify risk factors for weight-related problems in adolescence. Her team studied more than 2,500 boys and girls in middle school or high school and again five[22] and ten years later.[23] Girls who were dieting when the study began and also at the five-year follow-up had gained twice as much weight on average by the ten-year follow-up compared with girls who didn't diet at either time point, whether or not they were overweight to begin with. The initial characteristics that predicted weight gain during the study also predicted increases in disordered eating.

The strongest predictors were a history of dieting, concerns about weight, and weight-related teasing from family. Girls whose families nagged or teased them about their weight as young teenagers were twice as likely to gain weight over the next five years, compared with girls who weren't teased. One possible explanation is that the girls were already overweight, so they got teased, and then went on gaining weight. But that explanation doesn't work because the same relationship between dieting or teasing and weight gain was found for girls who started the study at a normal weight. Similarly, in another large, long-term study, children who dieted frequently at nine to fourteen years old were much more likely to move from normal weight to overweight over the next two years, by a factor of 4.8 in girls and 1.7 in boys.[24]

Body dissatisfaction also predicts weight gain in thin teenagers.[25] All junior-high and high-school students in one county of Norway were invited to participate in a study. Almost 1,200 of them filled out questionnaires both initially and at the follow-up eleven years later. The participants were at the low end of the normal weight range when the study began, but 74 percent of the girls and 26 percent of the boys felt overweight. Those students increased their weight and waist circumference more by early adulthood than their peers who correctly believed that their weight was fine.

As noted earlier, food restriction often leads people to binge. Indeed, teasing nearly doubled the risk of increased binge eating among girls in the Minnesota study, which was partially responsible for the weight gain.[26] Girls in the other large study who dieted frequently were twelve times more likely to report binge eating. Boys who dieted frequently were seven times more likely to binge eat. A third study of girls in northern California found similar results.[27] That study also showed that body dissatisfaction and controlled eating in young teens predict the onset of depression four years later, while body weight does not.[28]

What's a parent to do? By all means, talk to kids about healthy eating, but don't give them a hard time about their weight. Not only does that kind of talk contribute to eating disorders, but it's also likely to make them fatter in the long run. From an early age, allow children to stop eating when they're full, which almost all kids can do if parents provide a variety of healthy foods. Don't push them to finish what's on their plate, but don't allow them to make up any missed calories in candy later on. Teach them ways to deal with difficult emotions that don't involve eating, and set a good example with your own behavior. Encourage them to discover physical activities that they enjoy, especially those that they can continue into adulthood, and never talk about exercise as if it's a punishment for overeating. Many active children, especially girls, turn into couch potatoes in their early teenage years, a habit that is likely to persist throughout adulthood. Finally, if the local school or community is planning an obesity-prevention program, advocate turning it into a health-promotion program that doesn't focus on weight.

Adolescents may not be interested in advice from their parents, but younger children's dieting is strongly influenced by parental attitudes.[29] Girls whose mother or father encouraged them to diet at age nine were twice as likely to be dieting at age eleven—and eight times as likely if both parents promoted dieting. In contrast, for girls who started dieting in adolescence, parental encouragement did not influence their behavior. As we might expect by now, parental pressure to diet at age nine predicted greater weight gain by age fifteen, but only in the girls who went on diets. I find it hard to imagine that these young girls accurately estimated their risk of future weight gain when deciding whether to heed their parents' advice to diet, so reverse causality (girls who are more vulnerable to overeating are more likely to diet) seems unlikely to account for the results.

To avoid setting children up for failure, researchers from Project EAT recommend not talking about body weight or diets at home (for parents or children) and not using physical appearance to define anyone's identity. Children who see a parent skipping meals or expressing body dissatisfaction are likely to imitate that behavior, and parental encouragement to diet often produces unhealthy weight-control behavior and binge eating. Although you may not be able to change your feelings about your own body, try not to communicate negative thoughts to the kids. Instead, promote healthy habits by participating in sports or hiking as a family, making healthy snacks available, and bringing the family together for home-cooked meals on a regular basis.

The Dangers of Disordered Eating

Parents can also watch for signs of eating disorders. At age thirteen, 63 percent of girls in a long-term study in the United Kingdom were described by their parents as afraid of gaining weight or getting fat, and 11.5 percent as terrified.[30] Such concerns are a risk factor for the development of eating disorders. Because shame leads people to try to hide eating disorders, they are hard to diagnose. They are also hard to treat, with high rates of relapse, and they are dangerous. Anorexia patients have a sixfold higher risk of death, and bulimia patients a twofold higher risk. Among adolescents with bulimia, 53 percent have thought about suicide, and 35 percent have attempted it, compared with 3 percent of teenagers generally. Among those with binge-eating disorder, 34 percent have considered suicide, and 15 percent have attempted it.[31] Among anorexia patients, 31 percent have considered suicide, and 8 percent have made an attempt.

By the time I got my Ph.D., I had gained twenty pounds to reach my heaviest weight ever. After I started my first research job, I decided it was past time for another diet. This one got out of control. Once I lost that twenty pounds, I cut back my calories still further, to six hundred per day, started going to the gym for an hour every morning before work, and lost another thirty. My ribs were showing, but I wasn't satisfied. I still felt fat. By that time, my metabolism must have slowed down to almost nothing, because I stopped losing weight without changing my behavior. When I began to talk about needing to eat still less, my husband threatened to check me into a hospital. Or, he said, I could stop dieting. Forcing myself to start eating again was hard, with each bite requiring substantial effort. I'm lucky that I could do it on my own. Though I never qualified for a formal eating disorder diagnosis, I veered close to developing two of them.

The frightening thing about my story is how common it is. Dieting leads to pathological dieting in 35 percent of people. Among pathological dieters, 20 to 25 percent then progress to full or partial disordered eating.[32] The lifetime prevalence of full-blown anorexia in women may be as high as 4 percent, with bulimia and binge-eating disorder each at 2 percent.[33] In men, eating disorders are two to twelve times less prevalent, though they may also be underdiagnosed, as men are less likely to seek help.[34] Binge eating is the most common male eating disorder, with a lifetime prevalence of 1 to 3 percent, and boys who have to make weight for their sports have the highest risk of eating disorders.

In studies that follow participants over time, exposure to the ideal of thinness predicts later development of body dissatisfaction, which then predicts dieting, which then predicts eating disorders.[35] Dieting is the link between body dissatisfaction and eating disorders, as nondieters who are unhappy with their bodies tend not to develop eating disorders. All these risk factors have stronger effects in teenagers than in adults. Girls who were in the top 20 percent of body dissatisfaction at age thirteen were four times more likely to show disordered eating eight years later (24 percent) than those with lower body dissatisfaction (6 percent). Girls with depression and high body dissatisfaction were still more likely to develop eating disorders. Among the group with lower body dissatisfaction, high levels of dieting (top 12 percent) still predicted a 3.6-fold increased risk of eating disorders (18 percent versus 5 percent).

To break this cycle, Eric Stice of the University of Texas at Austin and his colleagues created the Body Project, an intervention that reduces the future risk of eating disorders in high-school and college students.[36] The program lasts three to four hours, and school staff can teach it effectively. The exercises reduce students' belief in the thin ideal by asking young women who are unhappy with their bodies to make arguments against the importance of thinness, in writing and in role play. By presenting these arguments to the group, the participants create a conflict between their behavior (arguing against the thin ideal) and their beliefs. Perhaps surprisingly, when people become aware of conflicts between their behavior and their beliefs, a problem that psychologists call cognitive dissonance, they often react by changing their beliefs. That is, we often figure out who we are by looking at what we do, instead of the other way around.

Indeed, the program works in just that way. Focusing on their cognitive dissonance motivates girls to reduce the value that they place on thinness. A randomized trial in an ethnically diverse group of high-school girls showed that three years later, participants had a 60 percent lower risk of developing disordered eating compared with their peers who participated in an expressive-writing program.[37] Reduced reliance on the thin ideal occurred first, followed by lower body dissatisfaction, fewer negative emotions, and less dieting. An online version of this program[38] also reduced the risk of eating disorders (though less effectively) and prevented the weight gain that occurred over the next two years in the control groups, in which girls read a brochure or watched a documentary on eating disorders.[39] By showing that an intervention that reduces dieting also prevents future weight gain, this randomized trial strengthens the argument that dieting is a cause of weight gain, not merely a correlated factor.

Many experts recommend dieting even though they understand the terrible odds against permanent weight loss, as the brain's energy-balance system fights to restore weight to the defended range. They figure it can't hurt to try, even if most people are unlikely to succeed. But dieting can and often does hurt. Its predictable results are temporary weight loss, followed by an elevated risk of self-hatred, eating disorders, and weight gain, especially in young women. No sensible person would want such outcomes, but that's what happens when people decide to pick a fight with their brains.

5

Our Brains Fight Weight Loss

A few decades ago, while trying to define humanity's future, researchers re-created some risks of our past. On September 26, 1991, four men and four women in futuristic jumpsuits entered Biosphere 2 with great media fanfare.[1] Glass and steel enclosed the three-acre compound, financed by $150 million from Ed Bass, the heir to a Texas oil fortune. The project was intended to demonstrate a self-sustaining, closed ecological system, as would be necessary if we were to colonize another planet. Bass was concerned that the first biosphere, the Earth, might not support life much longer, so he wanted to create an alternative.

From a spacefaring perspective, the experiment was disappointing. The plants did not transform enough carbon dioxide into oxygen, so managers activated the carbon-dioxide scrubber unit they had secretly installed before the experiment began. Biosphere's residents struggled to feed themselves from their half-acre garden, beset by insects and inadequate sunlight. Early in the experiment, their efforts produced only 1,780 calories per person to support a lifestyle that included three to four hours of manual labor per day, though the vegetarian food supplied enough nutrients. Things got so bad that the residents ended up eating some of their seed stocks. Six months into the experiment, they had lost between 9 and 24 percent of their body weight, an average of 15 percent, by dropping most of their body fat. With practice, they got better at agriculture and managed to grow enough food to maintain those lower weights for the final eighteen months.

Researchers took advantage of this accidental weight loss to study the effects of long-term food deprivation on metabolism.[2] A few weeks before they left the Biosphere enclosure, all the participants had suppressed metabolism and low thyroid hormones, and they conserved energy by avoiding spontaneous movement (that is, by not fidgeting). This experiment and others like it show that when we deprive our bodies of food, the brain fights back by changing the way the body uses energy.

Metabolism: The Brain's Secret Weapon

During the Biosphere experiment, the residents stayed hungry because they couldn't produce enough food. Those are the conditions that the body's energy-regulation system evolved to cope with—and that billions of people still confront today. Imagine if remaining underweight for two years had led the residents of Biosphere 2 to lower their defended weight range. In that case, instead of helping them recover once food became available, their brains would work to keep them from returning to their normal weights. As a response to a long famine, that wouldn't make any sense. The brain relentlessly applies the same logic to people who've been starving themselves voluntarily, even for years.

The brain's lasting metabolic compensation for weight changes was discovered by Rudy Leibel of Columbia University, who sounds like a New Yorker to the core. Straight shooting, charming, and profane, he's willing to fight for what he believes, which is lucky because this discovery was not popular. Puzzled by how inevitably people seemed to regain the weight they lost during the experiments with Jules Hirsch mentioned earlier, Leibel showed[3] that metabolism changes when people move outside their individual defended weight range, in either direction.[4] People whose body weight is 10 percent below their defended range typically burn 10 to 15 percent fewer calories per day than people who have the same weight within their defended range. These experiments are difficult and expensive, because they require the researchers to control food intake and exercise precisely for months. Leibel and his colleagues accomplish this feat by feeding their participants a liquid meal replacement while they live in an inpatient facility and maintain their regular exercise routine throughout the study.

Leibel finds that metabolic suppression persists in dieters who have kept weight off for one to six years,[5] so he scoffs at claims that the occasional weight-loss success story disproves his ideas. "If you talk to people who've done it—not the studies, but people who have actually lost weight and managed to keep it off—they'll tell you what I'm telling you," he says: that the only way to achieve this goal was to allow themselves to be hungry all the time while increasing their physical activity substantially. Indeed, his point is supported by data on the eating and exercise habits of people listed in the National Weight Control Registry, who have lost at least thirty pounds and kept it off for one year. A calorie calculator says that Dennis Asbury should have needed 2,100 calories a day to maintain his weight at 150 pounds, but instead he found that he needed to eat 400 to 500 calories less than that. Such metabolic suppression is the difference between being within the defended weight range and being below it. Many people blame others for eating too much or exercising too little, assuming incorrectly that both are under full voluntary control, but it's much harder to justify holding people responsible for diet-induced changes in the way the body burns energy.

Metabolism is often discussed as if it's magic, to paper over whatever the speaker doesn't understand about the response to diet and exercise. The term has a scientific definition, though, as a measurable property of the body. When food is digested, the body can do three things with the energy: use it to do work (like lifting heavy objects), store it as fat, or turn it into heat. By far the most energy turns into heat. Some of that heat maintains body temperature, and the rest is released into the world because digestion is an inefficient way to turn food into usable energy.

In the technical sense, metabolism is simply how much heat the body produces. It's measured in calories, the kind that are listed on food labels, by placing a person in an insulated room with water-filled walls and then measuring the change in water temperature. Metabolic rate can be estimated more cheaply from oxygen consumption by making some assumptions about the person's diet. Thyroid hormones provide the main control of metabolism, but it's also influenced by adrenaline released during the fight-or-flight response and by muscle activity, including shivering with cold or tensing up from anxiety.

A similar compensation happens in the opposite direction when people eat too much: their bodies burn more energy for a while. That's why people often wake up feeling too hot a few hours after a big meal. This reaction has frustrated experimental attempts to induce weight gain in thin people.[6] In the 1970s, a group of prisoners agreed to gain weight for a research study. Although they eventually managed it, some of them had to eat as much as ten thousand calories a day for four to six months to increase their weight by 20 to 25 percent. At that point, their metabolic rates had increased so that they were burning 50 percent more energy. The researchers then compared the now-overweight prisoners to obese people who had dieted down from a higher weight. To maintain the same weight, the successful dieters had to eat about half as much as the overfed prisoners because of differences in metabolism between the two groups. However, the response to overeating is less persistent than the response to starvation. In both rodents and people, the initial resistance to weight gain seems to fade after a few months.

Contestants from *The Biggest Loser* try to win a cash prize by losing a massive amount of weight through calorie restriction, heavy exercise, and letting trainers yell at them for our viewing pleasure. Although this approach does cause fat loss while sparing muscle, the dieters show strong metabolic suppression, amounting to five hundred calories per day, after losing an average of 38 percent of their body weight, even while they remain (barely) obese.[7] The participants who lose the most weight also show the most suppressed metabolism.

When metabolism declines after weight loss, the major change is not in how much energy muscles burn at rest but in the energy cost of mild physical activity, like walking across a parking lot. To reduce metabolism, the energy-balance system increases the efficiency of skeletal muscle—that is, the amount of force produced per calorie burned—by 15 to 20 percent. In other words, dieters who've lost a lot of weight get better fuel efficiency when they move around. During a famine, it's useful to be able to walk farther to search for food without using more energy. For dieters, though, needing to exercise longer to burn the same number of calories is not such good news. Worse yet, this low metabolism lasts at least six or seven years after weight loss, which is the longest that anyone has measured it—and probably forever.

The Brain's Weight Regulator

Anyone who believes that an obese person could become thin simply by following the lifestyle of a thin person is ignoring the power of the energy-balance system or "weight thermostat" centered in the hypothalamus. This system directly controls hunger, physical activity, and metabolism, based on information from the body about its short- and long-term needs. As the command center for weight regulation, the hypothalamus receives dispatches from other parts of the body, carrying information about the availability of calories, as well as specific nutrients. It uses this information to figure out what we should eat and then nudges us with unconscious signals to make it happen.

To decide how much we should eat at a particular meal, the energy-balance system relies on signals released in the gut, which then activate the vagus nerve to send information to the brain. A messenger called ghrelin is released before meals and stimulates hunger. When we've had enough food, the gut releases satiety signals, including a peptide called cholecystokinin, along with many others, which tell the brain that it's time to stop eating. As a result, we feel less hungry, and food becomes less stimulating to the brain's reward system.

Thus one of the ways that satiety signals cause us to stop eating is by making food less enjoyable. By paying close attention to the difference in satisfaction between the first bite of food and the last, you may be able to demonstrate this effect for yourself. Although these signals determine how much an animal eats at one meal, they don't regulate body weight. If free-feeding rats are given an injection of cholecystokinin at the beginning of every meal, they eat less food per meal. They don't lose weight, though, because they eat more meals to compensate, so they end up taking in the same number of calories.[8] The regulation works the same way in people.

A lot of what we know about the control of body weight comes from studies of rodents, whose eating and weight regulation are a lot like people's. Mice and rats are omnivores, who like the same foods that we do. Their body weight is regulated in the same way, with weak protection against weight gain and strong protection against weight loss. When given access to tasty food, rodents gain different amounts of weight, depending on their individual genes. All the genes that influence weight in people also affect weight in mice. Under stress, rodents are more inclined to eat sweet and fatty foods, and repeated weight loss can lead them to binge on Oreo cookies, as we learned earlier. Finally, like us, both wild and laboratory rodents have gotten fatter over the past few decades, suggesting that whatever environmental changes are responsible for the increase in human obesity are probably affecting rodents in the same way.

In rodents and people, body weight is regulated by feedback signals from leptin, which tells the brain whether current fat stores are on target for the defended weight range. That's a body-fat range rather than a body-weight range. That is, this system doesn't care about how muscular a person is or what number the scale reads, just how much energy is stored as fat. When leptin levels in the blood decline, as people lose enough fat to drop below their defended range, the brain's starvation alarm goes off. To save the life of a starving person (or so the brain assumes), the hypothalamus reduces metabolism, decreases physical activity, and increases hunger. These symptoms can be reversed by injections of leptin, which tricks the brain into thinking the weight has come back, so the emergency is over. We know that these actions of leptin occur in the hypothalamus because removing the protein that receives signals from leptin in this area alone leads to massive overeating and obesity in rodents by fooling the brain into thinking that body fat is at zero.[9]

The opposite effect occurs when small doses of leptin are given to people who've lost a lot of weight. Most of the starvation-like symptoms of weight loss are abolished, including persistent hunger and reduced metabolism, as leptin tricks the energy-regulation system into believing that the body has enough fat.[10] So far, that approach has been used only experimentally, in part because leptin is broken down by the digestive system, so it has to be delivered by injection. The more challenging issue is that the FDA standards for drug approval require that a drug must help promote weight loss and don't recognize the prevention of regain as a valid goal. Thus pharmaceutical companies don't have an incentive to do the clinical trials required to get such drugs licensed for use.

Similar, but smaller, effects on weight regulation come from insulin, which the pancreas produces in response to elevated blood sugar. Insulin's main job is to direct the body to store energy when blood sugar is too high and release energy when blood sugar is too low. In addition, insulin acts on the hypothalamus to reduce food intake, but less strongly than leptin. Extra insulin by itself causes only a small amount of weight loss.

Both leptin and insulin increase the brain's responsiveness to satiety signals, so that when leptin is high (usually because there's a lot of fat in the body), people stop eating sooner. In this way, the long-term goals of the energy-balance system—to make people lose or gain weight to reach their defended range—are reflected in their short-term behavior, as the brain guides them to eat less or more than usual. Under most circumstances, this system is extremely effective in maintaining a stable weight, despite short-term fluctuations in food availability.

The brain's energy-balance system works beneath conscious awareness. As with breathing, we can control the balance of eating and exercise for a little while, but we don't need to think about it to make it happen. "I don't know how to do the majority of things that keep me alive," says Dennis Asbury. "I don't know how to digest food. I don't know how to convert fat back into energy. I don't know how to get glucose across the blood-brain barrier. You know, any of the stuff that makes me work. I'm just oblivious to it all. It's automatic." He's convinced himself that mindful eating is worth a try, as rigidly controlling his eating and exercise hasn't worked out so well for him. As a first step, he tried going a week without counting calories or weighing himself, just eating healthy food according to his hunger signals. To his surprise, his weight went down half a pound, calming his anxiety that if he stops counting and weighing, his eating will end up out of control. "I don't have to count calories," he reports. "My body is counting them. It does the math with perfection."

Soon after I quit dieting, I tried a similar experiment on myself. For a month, I ate when I was hungry and stopped when I was full. I also tracked everything I ate, as I used to do when dieting. The results were illuminating. My calorie intake varied from 1,400 to 2,400 calories on any given day, but over any week, I averaged 1,800 calories per day. The days when I ate a lot, usually because of a party or other special occasion, were typically followed by two or three days when I was less hungry and ate less. As an experienced dieter, I was impressed to realize that my hypothalamus is much better at keeping track of calories than my deliberate efforts ever were. Now I feel like I wasted all the time and energy that I spent planning, weighing, measuring, and tracking my food. Since then, I've been comfortable with letting my internal cues do the work that I used to feel was my responsibility, though I do help my brain get it right by eating whole, unprocessed food.

The Defended Range Creeps Upward

The puzzle isn't how the energy-balance system works—scientists understand that pretty well—but why it allows some people's defended weight range to creep up over time. If we could prevent or reverse that process, then diets would work for most people. Weight loss and its maintenance probably would never be effortless in a world of constant temptations to the reward system, but they might become as easy as the average online commenter believes, requiring a modest and achievable application of willpower. The inventor of a drug that lowers the defended range could make a fortune, though it's not clear that teenage girls (or many adults) in our culture could be trusted with the ability to become as thin as they desired.

One possible explanation is that in an emergency, those careful calculations go out the window. The brain is an especially vulnerable part of the body because the only fuel that it can use is glucose, the main sugar in blood. To prevent major damage to the brain caused by a lack of fuel, which can happen in as little as five minutes, a set of circuits called the emergency feeding system can override long-term energy goals. Low blood sugar (say, if we forget to eat lunch) can activate a cascade of stress responses that reduce insulin release and increase stress hormones. These signals reduce the use of glucose and increase its production, in an attempt to restore blood sugar to normal levels. When the emergency feeding system is active, the drive to eat is intense, regardless of how much food we've had recently or how much leptin is in our blood. Mike Schwartz, a researcher at the University of Washington, suspects that repeated activation of the emergency feeding system causes many cases of obesity and diabetes.

Researchers are starting to get a grip on how the defended range increases, though many details remain to be filled in. Its cornerstone is leptin resistance, the weakening of neural responsiveness to leptin. This impaired sensitivity is a common feature of all forms of obesity, caused by anything from eating too many cookies to genetic mutations that damage the energy-balance system. People with leptin resistance have a lot of leptin in their blood, but their brains act as though they're not hearing the signal. This problem explains why doctors can't make obese people lose weight simply by giving them extra leptin—an idea that caused a lot of excitement when leptin was first identified in 1994—because their brains have stopped responding normally to the hormone. A similar loss of responsiveness to insulin underlies type 2 diabetes. Recent work suggests that leptin resistance is not just a consequence of obesity, as researchers thought for many years, but may also be its cause.

No one knows for certain what causes leptin resistance, but the strongest current possibility is that a process called cellular inflammation may damage neurons in the hypothalamus. Cellular inflammation is a local immune response, often caused by a physiological imbalance. When researchers cause inflammation in the hypothalamus, rodents overeat, gain weight, and become less responsive to insulin and leptin, as well as more susceptible to diet-induced obesity.[11] Blocking the chemical cascade that leads to inflammation in the hypothalamus has the opposite effects. One report suggests that long-term exposure to a high-fat diet leads to the death of neurons that inhibit eating in the mouse hypothalamus, which could permanently tilt the balance in favor of overeating, increasing the likelihood of further weight gain.[12] Differences in inflammation before mice start eating a high-fat diet seem to explain strain-specific tendencies toward obesity,[13] and inflammatory molecules in the blood predict the likelihood that individual people will develop diabetes.[14]

Inflammation seems to be an early step on the pathway that leads to a higher defended weight range. If inflammation in the hypothalamus causes leptin resistance, then the inflammation should occur before leptin resistance develops. In rats bred for susceptibility to diet-induced obesity, that's what happens.[15] Within twenty-four hours of starting a high-fat diet, these rats have inflammatory markers in their hypothalamus, which continue to increase through day three, stabilize for two to three weeks, and then increase again—following the same pattern as the rats' increased food intake. This lockstep progression suggests that inflammation increases the defended range, leading to more fat storage, which further increases the defended range, in an ongoing cycle. Inflammation is widespread in obese people as well, but we don't yet know whether it precedes weight gain.

Leptin-resistant animals show many symptoms of the emergency feeding response, including high blood sugar. Because mice that lack leptin show symptoms of diabetes, some researchers propose that the brain may use the emergency feeding system to respond to long-term fuel deficits signaled by low leptin in addition to short-term energy deficits. Putting leptin into the brain restores normal blood sugar in both diabetic and leptin-lacking rodents, lending support to this hypothesis. This system must be at least partially separate from the system that controls energy balance, because the emergency-feeding response still occurs after damage to the main energy-balance area of the hypothalamus, which is called the arcuate nucleus.

The arcuate nucleus contains neurons that stimulate eating and others that inhibit it. The stimulatory neurons are activated by ghrelin, the short-term hunger signal from the gut, and inhibited by leptin and insulin. Activation of these hunger neurons causes lab animals to gobble down food, no matter how much they've eaten recently, and makes them more willing to work for food.[16,17] The more hunger neurons become active, the more animals want to eat. Weight loss activates these neurons, which increases hunger—sometimes to ravenous levels in people who've lost a lot of weight. The neurons that inhibit eating are activated by leptin and reduce feeding over a slightly longer timescale, closer to a day.

Together, these neural circuits sense the body's nutrient and energy needs and adjust eating, exercise, and metabolism accordingly. Although we can modify these decisions to some extent through deliberate effort, at least for a while, it's not clear why we would want to do that. Left to its own devices, the energy-balance system can maintain a stable weight for most people more effectively than deliberate attempts to calculate our calorie needs and intake.

The Real Meaning of Personal Responsibility

Diet plans ask us to take "personal responsibility" for our weight by following a rigid set of rules developed for an average person and designed to make everyone's body look the same. Instead, let's take responsibility for creating lives that are a good fit for our own individual bodies. Some researchers argue that we shouldn't tell the public the truth about the difficulties of weight loss because it might discourage dieters, including the few who do succeed. That philosophy discounts the enormous costs borne by the majority who repeatedly regain their lost weight and then feel like failures. It also disregards reality. We'd feel fine about discouraging people from stretching their legs to make themselves taller. For many of us, dieting is barely more successful in the long term.

The diet industry has adopted a strategy of taking credit for weight loss and then blaming the predictable regain on individual lack of willpower when it actually results from the brain's energy-balance system working correctly to reverse weight loss. If the industry acknowledged that increased hunger, decreased metabolism, and intense psychological distress are normal responses to substantial weight loss, clinics would face a huge marketing disadvantage. So they don't. Ignoring the problem, of course, doesn't make it go away, but it does allow diet clinics to profit from a steady supply of repeat customers and to sell lifetime memberships with a straight face. A better question is why so many of us put ourselves through this process again and again. We suffer through hunger and puff away at the treadmill, only to see our efforts wasted as our brains win another round in this long struggle, then meekly accept the judgment that this outcome is our own fault.

Dieting teaches us to ignore the body's signals of hunger, which leads most people to overeat when the opportunity arises. Long-term deprivation causes binge eating in susceptible people, thus causing the weight gain that diets were intended to cure. Choosing instead to practice mindful eating puts the energy-balance system back in control, freeing us to focus our willpower on more important priorities. There are a variety of research-tested ways to do that, but first let's take a look at the factors that lead us to gain weight.

Part Two

Why We Gain Weight

6

Early Life and Adult Weight

The winter of 1944–45 was unusually cold in northern Europe, but the news was good. The Allied armies rapidly gained ground, and the war seemed likely to end soon. In September and October, soldiers liberated the southern provinces of the Netherlands from a Nazi occupation that had begun in 1940. To support the Allied attack, the Dutch government-in-exile called for a railway strike, and thirty thousand workers complied. An attempted airborne landing failed to free the western provinces, though, as the German army used its own trains to bring in troops, and the Allies moved on toward Berlin.

In retaliation for the Dutch resistance, a German blockade prevented the delivery of supplies to the parts of the country that remained under Nazi control, so civilians had to scrap for calories and heat any way they could manage. Rations dropped from 1,200 calories per day in November to 800 in January to fewer than 600 in February, mostly bread and potatoes. Earlier in the war, pregnant and nursing mothers had gotten supplemental rations, but by late fall there was not enough food available to continue them. People burned furniture and ate tulip bulbs. Farmers put up signs saying that they would no longer trade food for the fancy linens of the city folk. More than twenty thousand people died between November 1944 and May 1945, and everyone starved. The biological consequences of the Dutch Hunger Winter would echo through the rest of the survivors' lives, their children's lives, and in some cases their grandchildren's lives.

These war victims contributed a great deal to our understanding of the lasting effects of early life experience. Centralized health care and excellent record keeping allowed researchers to study the long-term effects of a single period of malnutrition of known length and timing during an otherwise adequately fed lifetime. They found that a baby's nutrition in the womb had lifelong effects on weight.[1] When women starved in the last few months of pregnancy, their babies were born small and remained small and thin for the rest of their lives, despite having access to normal amounts of food. The grandchildren of these women also had low birth weights. In contrast, if women starved in the first half of their pregnancies and then ate well, their babies had normal birth weights but were heavier as adults. When mothers starved early in gestation, their daughters (but not sons) had strong odds of increased weight and waist circumference as adults. These women were also more susceptible to diabetes, regardless of their adult weight. The same pattern was found in women whose mothers were pregnant during the massive famine caused by Mao's Great Leap Forward in China, compared with those born in areas where food remained available.[2] No one knows for sure why male babies are less affected by early starvation, but one possibility is that weight regulation is more important for women because it affects their fertility.

Pregnancy Prepares Babies for Their World

A relationship between the conditions during fetal development and adult characteristics makes sense if we think about pregnancy as an opportunity to match each baby to the external environment in which he or she will live.[3] In people and other animals, biology uses information from the environment to shape the growth of babies before they're born. If food is scarce out there, a smaller, thinner body will make it easier to get enough to eat. Tastes from the mother's diet that reach the baby's tongue in the womb signal what is safe to eat around here and shape the child's later food preferences. If danger lurks all around, a hair-trigger stress system that reacts to any hint of trouble may be necessary for survival, so the health costs of long-term stress could be worth paying. If conditions are welcoming, though, then people will be most successful if they take advantage of the abundant resources to grow big and strong, rather than wasting energy jumping at shadows. When the mother's environment during pregnancy correctly predicts the world that the baby will live in, then this developmental plasticity makes the baby more successful. Problems can result, though—either because of inherent trade-offs, as in stress responsiveness, or because the adult ends up in an environment that doesn't match the one to which its young body was adapted.

A common developmental mismatch occurs when people who are born into deprivation then grow up with abundant food. In rich countries, the odds of abdominal obesity (the most dangerous type for health), heart disease, diabetes, and high blood pressure are all higher in people who had low birth weights than in those who were heavier as newborns. Populations that move directly out of poverty to a diet of processed foods high in refined grains, sugar, and fat have an especially high risk of diabetes.[4] A similar effect is seen in children whose mothers were deprived of protein during pregnancy. On the other hand, people who had high birth weights generally need more food as adults and don't do well during famines. In an environment with abundant food, both underweight and overweight babies tend to grow up to be obese. This is true not only for people but also for laboratory animals, and the availability of tasty food later in life makes all these odds higher.

Modern babies are heavier than previous generations by six months of age, long before willpower, diet, or exercise choices become an issue.[5] That increase began in the 1970s in many countries, including Norway, Sweden, Denmark, Canada, and to a lesser extent the United States. Larger babies tend to grow into larger adults, so increases in obesity within a population can be seen earliest in babies, then in children, and finally in adults.[6] Similarly, babies who grow more quickly are more likely to become obese as adults. Among children who are overweight at ages ten to fifteen, 80 percent will be obese by age twenty-five.[7] These observations turn our obsession with individual responsibility for food and exercise choices on its head, raising the question of what biological changes predispose people to gain weight from the earliest stages of their lives. An increase in the brain's defended weight range due to the abundance of available food seems like a reasonable explanation.

In support of this idea, babies are more likely to have high birth weights if their mothers gained a lot of weight during pregnancy. Because babies are almost always genetically related to their mothers, any correlation between their weights might be genetic as well. But a clever study solved this problem by comparing siblings after the same mother gained different amounts of weight in different pregnancies.[8] On average, the baby's weight increased steadily with the mother's weight gain, with each ten-pound gain predicting an additional 1.2 ounces of birth weight. Gaining fifty pounds roughly doubled the likelihood of having a high-birth-weight baby.

Researchers are beginning to figure out how maternal obesity influences fetal brain development. Most of this work is done in rodents, because the ethical bar is high for experimenting on human babies (as it should be). In young animals, leptin is necessary to develop normal wiring in the brain circuits that control eating.[9] As adolescents, mice that lack leptin have a much lower density of the fibers (called axons) that carry signals from the arcuate nucleus to other parts of the hypothalamus. Injecting these mice with leptin causes these axons to grow normally, but only in early life. By adulthood, the axons no longer grow in response to leptin, although leptin can still influence their synapses, the connections between neurons. When mouse mothers are fed a high-fat diet during breast-feeding, pups show the same reduction in axons from the arcuate nucleus, as well as impaired blood-sugar regulation.[10] Both problems are prevented if researchers block the action of insulin in arcuate neurons, suggesting that excess insulin in the pup's brain interferes with development of the hypothalamus.

Normally, young rodents experience a surge of leptin in the second week after birth, which seems to be important for brain development. Rat mothers with obesity caused by a high-fat diet have pups that show an exaggerated version of this surge, which releases more leptin and lasts longer than normal.[11] At one month of age, when their weight is still normal, leptin in the arcuate nucleus of the hypothalamus has less ability to inhibit eating, which persists into adulthood. That is, these lean adolescent animals already have leptin resistance, preventing the brain's energy-regulation system from realizing how much fat their bodies carry. This leptin resistance may explain why their weight increases later in life. Mice born to diabetic mothers show a similar set of changes.[12] Injecting rodents with leptin early in life can also lead to leptin resistance in adulthood. In people, mothers with diabetes or obesity have babies with more leptin and insulin in their blood at birth, but the functional significance of this change is not yet clear.[13]

Leptin also regulates the birth of new neurons in the developing hypothalamus, the developing cortex, and the adult hippocampus, a region important for memory and spatial navigation. Mice lacking leptin have a lower density of neurons in the hypothalamus than normal mice. Feeding a rat mother a high-fat diet during pregnancy and lactation increases the production of neurons in the pup's hypothalamus that contain orexin and melanin-concentrating hormone, both of which act to increase food intake. In adult mice, eating a high-fat diet quadruples the number of new neurons born in the hippocampus.[14] Blocking the birth of new neurons in this region leads mice to gain much less weight on the high-fat diet, suggesting that these new neurons make animals susceptible to diet-induced obesity.

Stressed Children Become Heavy Adults

Another early-life experience that increases the odds of adult obesity is severe childhood stress, which can be caused by poverty or abuse. In many species, including people, early life stress permanently makes the stress response system more reactive, which helps animals survive in a threatening environment. The brain changes caused by early stress make stress hormone responses easier to activate and harder to turn off. One cause of this exaggerated response is a decline in the number of stress hormone receptors available to receive the feedback that shuts the response down. The biological basis for these changes is well understood from research in rodents and monkeys, and the broad outlines appear similar in people.

The persistent stress hormone exposure resulting from this hypersensitive stress system has serious drawbacks. It increases vulnerability to abdominal obesity and both physical and mental health problems. For people who grow up in a war zone, where extreme vigilance is needed just to get through the day, those trade-offs may be worth it, but the price is high.

Childhood abuse or neglect predicts adult obesity in people. One study found that frequent physical abuse increased the rate of adult obesity by 71 percent, sexual abuse by 40 percent, and frequent fear by 34 percent.[15] The authors estimate that 8 percent of obese adults and 17 percent of the severely obese owe their weight to childhood abuse. Other researchers found that normal-weight children whose parents gave them little or no support, as perceived by their teachers, were seven times more likely to be overweight in young adulthood than their peers.[16] In the same group of children, those who were neglected were more than eight times as likely to become obese by young adulthood.[17] A meta-analysis of forty studies with more than 190,000 participants concluded that childhood mistreatment was associated with a 36 percent increase in the incidence of adult obesity, which typically developed after adolescence.[18] Emotional neglect did not have a clear effect on body weight in that analysis. Studies that track participants over time tend to find that the abuse preceded and predicted the weight gain, rather than being a consequence of it.

Maternal Smoking and Weight

Although nicotine suppresses appetite in the short term, nicotine exposure before birth can increase the long-term odds of obesity. Strong evidence suggests that babies whose mothers smoked during pregnancy are 50 percent more likely to become obese later in life.[19] Since public health campaigns have reduced smoking over the past few decades, women who smoke while pregnant now differ from women who don't, which complicates the interpretation of studies. Pregnant smokers are poorer, less educated, heavier themselves, and less likely to breast-feed, which are all qualities that predict childhood obesity. Yet adjusting statistically for these characteristics does not remove the risk posed by smoking. Similar risks were found in older U.S. studies and in European studies, in which smokers were less different from the rest of society. In addition, mothers who smoke after pregnancy do not increase their children's odds of obesity, although they are similar to women who smoke during pregnancy. Finally, this association is supported by laboratory animal studies, which find that prenatal nicotine leads to lower birth weight and increased body fat later in life. For all those reasons, the association between smoking during pregnancy and later obesity is probably real.

Researchers do not know how nicotine influences body weight, but it reduces the defended range in adult smokers, so it might help to determine the initial defended body size of babies. A long-term study of Australian twins showed that smoking during pregnancy leads to lasting chemical modifications to DNA, perhaps explaining why the effects are lifelong.[20] In rats, low doses of nicotine during pregnancy increase the production of neurons that contain orexin and melanin-concentrating hormone, both chemical signals that stimulate feeding, in the lateral hypothalamus.[21] It also increases the production of neurons containing enkephalin, a natural pain reliever related to morphine, in the paraventricular nucleus of the hypothalamus and in the amygdala. In normal adults, activation of these neurons increases the animal's desire for high-fat food. During adolescence, nicotine-treated offspring showing these brain changes eat more high-fat food and consume more nicotine or ethanol than untreated rats, without any tendency to overeat on regular rat chow, suggesting that their brains are more sensitive to rewards. If similar changes occur in people, that could explain the increased likelihood of obesity in children whose mothers smoked during their gestation.

Environmental Chemicals and Obesity

If the recent rise in obesity is due to an epidemic of poor willpower, then our pet cats and dogs need to work on their self-control. Not to mention our laboratory animals and wild rats. Like people, all these animals have gotten fatter over the past few decades, according to research on more than twenty thousand animals from twelve separate populations, all living with or around people.[22] The laboratory animals are particularly interesting, as their diets and living conditions were recorded and in most cases have not changed. Nonetheless, they're choosing to eat a lot more of the same old chow than previous generations. Macaque monkeys have increased their weight by 7.7 percent per decade, which more than doubled their rate of obesity, and marmosets showed similar increases. Chimpanzees increased their weight by more than a third and their odds of obesity more than tenfold. Even wild rats increased their likelihood of obesity by 20 to 25 percent per decade, whether they lived in the city or the country. Across species, females had more substantially increased odds of obesity than males did. The effect was stronger in laboratory animals than in free-living animals, perhaps because owners heed the vet's recommendation to reduce their pet's food intake to prevent obesity. It's hard to attribute the weight gain of these animals to laziness, gluttony, or the increased availability of convenience stores or car travel. Instead we need to look for causes that are widespread in the environment. Like industrial chemicals, for instance.

Prenatal exposure to a variety of environmental chemicals has been linked to obesity, at least in laboratory animals.[23] A common ingredient in PVC plastics increases the differentiation of fat cells and increases body fat in young rodents,[24] but it's not clear if it has the same effects in people. The organophosphate pesticides increase the rate of obesity and poor health in rodents, but again their effects in people are unknown. A widespread ingredient in household plastics like water bottles, whose chemical name is abbreviated as BPA, can activate estrogen receptors, which has led to concerns because almost everyone these days—fetuses included—has it in their bodies. Animal studies indicate that BPA can increase insulin release and the development of fat cells,[25] but so far the studies on people have been weak.

Researchers have typically measured exposure only at one time point, well after birth. In addition, it's a tough statistical problem to prove a correlation between weight gain, which is common in our society, and BPA exposure, which is also widespread. Still, the evidence is suggestive enough that pregnant women and small children might want to avoid plastics if weight is a concern.

Some lasting chemical modifications to DNA due to prenatal influences can be passed down through one to three generations. For example, some effects of the Dutch Hunger Winter were found in the grandchildren of those originally affected. The effects on the children lasted a lifetime. Six decades after being starved in utero, the survivors still had chemical modifications on their DNA that are consistent with transgenerational transmission. In particular, they had less methylation on the gene for insulin-like growth factor II than their unaffected siblings, but only if they had been in their first trimester of development during the famine.[26] Adults who were in the third trimester at that time showed no gene modifications. In laboratory animals such gene modifications, when carried in sperm or eggs, are one way that environmental influences are passed down to future generations.

In some cases, effects of chemical exposure become apparent only in later generations. The pesticide DDT is banned in the United States but used frequently in developing countries to control the mosquitoes that cause malaria. When fetal rats are exposed to this pesticide, it does not influence their own weight. However, their exposure leads to obesity in more than half of their grandchildren.[27] This effect is probably caused by gene modification because both mothers and fathers transmit it, and because changes in DNA methylation consistent with this idea are found in the obese generation.

Diet and Weight Across Generations

The effects of a high-fat diet can also be transmitted to the offspring of the rodent generation exposed in the womb.[28] Again, both mothers and fathers can transmit this inheritance, suggesting that the effect occurs through DNA modification. Both the children and grandchildren of mouse mothers that ate a high-fat diet during pregnancy show increased body length and reduced insulin sensitivity, a precursor of diabetes. Only the children, though, gain extra weight—and they do it without eating more. Instead their energy use is more efficient. Many other rodent studies support the idea that metabolic abnormalities related to diabetes due to over- or underfeeding during pregnancy can be transmitted through multiple generations.

Historical records suggest that a similar process happens in people. Swedish parish records of births, marriages, and deaths go back to the 1800s. Grain prices, records from community meetings, and regional harvest statistics indicate how much food was available in different years. By combining these sources of information, researchers traced the life trajectory of many individuals, their parents, and their grandparents. By the parents' generation, crop failures no longer led to food shortages, as modern transportation made it possible to bring in food from other parts of the country, so the researchers could examine the lasting effects through multiple generations of food deprivation that occurred only in the grandparents. Many such studies found that a grandparent's nutrition just before puberty, during the so-called slow growth period that lasts from ages eight to ten in girls and nine to twelve in boys, is particularly important for the grandchildren's later health. If the paternal grandfather had good access to food (no crop failures) during the slow growth period, his grandchildren had a fourfold higher risk of diabetes and a higher rate of cardiovascular disease.[29] Similarly, a study in South India showed that low birth weight in either parent predicted that their adult children would have a higher risk of metabolic syndrome, a constellation of symptoms (high waist circumference, blood pressure, blood sugar, and cholesterol) that is a risk factor for serious health problems like diabetes and heart attacks.[30]

Taken together, this research suggests that many experiences during fetal and early childhood development can affect weight regulation into adulthood, probably by adjusting the defended weight range of the brain's energy-regulation system early in life. Community conditions or parents' life experiences can increase young children's likelihood of becoming obese adults long before they have any say in the matter. In addition, eating too much or exercising too little may not be the only culprits in obesity. People who reported identical eating and exercise habits had heavier bodies in 2006 than in 1988.[31] That finding suggests factors that influence metabolism—such as environmental chemicals, gut bacteria, or prescription drugs—may be important contributors to weight gain.

Yet we blame, bully, and punish children who are overweight. Lecturing kindergartners on the need for dietary restraint has no chance of reducing obesity, but it's likely to cause psychological damage and increase future weight gain. In extreme cases, obese children have been taken from their parents' custody and forced onto diets or into weight-loss surgery. Diets have no better lasting effectiveness in children than they do in adults, though they can interfere with growth. In any case, severe obesity in children often results from a genetic mutation that causes a malfunction in the brain's energy-balance system, a problem that parents can't be expected to fix. The urge to blame parents for their children's weight is part of the impulse to blame adults for their own obesity. This cultural pattern is a source of lasting stress for parents and children alike. And that, as you may recall, leads to weight gain.

7

Stress, Shame, and Stigma

Sylvia Berman never worried about her weight until a series of unfortunate events changed her body and her life. She was thin and athletic, identifying herself as a dancer, runner, swimmer, skier, and hiker. Being able to trust her body's physical abilities was important to her, and in turn she felt like she could sense what it needed for nutrition and care. When her son was born, she quit her job as the manager of a research laboratory, but she remained interested in science. She loved explaining the world to her children and their friends, like the day she told their class how rainbows form.

After her daughter's birth in 1985, Sylvia got an infection and was rushed back to the hospital with sepsis, a whole-body inflammatory response that can cause rapid organ failure and indeed almost killed her. She was so sick that her mother asked the church to send someone over to pray for her. To control the sepsis, her doctors gave her powerful antibiotics. Because she kept getting infection after infection once they sent her home, she stayed on oral antibiotics for months. That treatment saved her life, but it also most likely devastated her gut bacteria, which strongly influence body weight.

Soon afterward, her doctors discovered a swollen area on her thyroid gland, which produces hormones that control metabolism. That's a symptom of hypothyroidism, a condition that slows metabolism, leading to fatigue, cold sensitivity, and slow weight gain over a period of years. In most people, the problem has been happening for a while before it's diagnosed. Her doctors put her on steroids, though she can't remember why, and she gained about fifty pounds, even though she had started running again.

The final blow came as she was finally getting over her illnesses. While chasing after a friend's child on a winter outing to Lake Michigan, she slipped on the ice and fell, shattering her ankle. A talented surgeon put it back together with metal pins, but it's never been quite right since. To this day, she walks with a cane. Her exercise routine ended abruptly, and her weight continued to increase. Within six months, she'd gained another fifty pounds or more.

People treated her differently. She noticed that right away. Once she walked into a Neiman Marcus department store to buy a present for her mother-in-law. The last time she'd been there, when she was thin, she'd enjoyed the experience. This time she stood at the counter for a long time waiting for the same salesclerk to stop ignoring her: "The bigger you get, the more invisible you become." When she tried swimming again, children made loud comments about her size. After it happened again, she stopped going to the pool. The stress wasn't worth the benefit anymore. At times, stress seemed to be taking over every aspect of her life. The comments from strangers in the grocery store began to wear her down.

Although she tried every plausible approach over three decades to lose the weight, her brain had started defending the higher range, and nothing seemed to work. Plenty of people had suggestions, but she soon tired of hearing the same ideas over and over. She feels ashamed that she can't lose weight, though Qigong satisfies her need to exercise and helps her stay healthy. Even knowing everything that happened to change her body, she can't shake her sense of responsibility for failing to become thin again. Like many of us, Sylvia finds her weight to be an enormous source of stress. When I ask whether she could consider accepting her body as it is, she snaps, "I'm never going to say this is okay. This is not okay."

Stress Makes Us Sick

The stress system is adapted to respond to immediate, short-lasting threats like lion attacks, not to long-term problems like poverty, a bad marriage, or an unfair boss. An effective hormonal stress response starts quickly and ends quickly, lasting minutes to hours. In people, the hypothalamus responds to stress by ordering the release of a hormone called cortisol from the adrenal glands. Its job is to get the person out of a life-threatening emergency, without considering the long-term consequences.

Cortisol's actions on the body contribute to immediate survival by increasing alertness, improving the ability to think, and making sure that enough energy is available to muscles for running or fighting. This hormone also directs resources away from long-term concerns by inhibiting growth, repair, reproduction, digestion, and immune responses. That's why lasting stress is damaging, because the body can't afford to shut down those processes for too long. If cortisol stays high, it can make infections hard to fight off and damage parts of the brain that are important for memory and emotion. To keep the stress response brief, cortisol in the blood provides feedback to the brain that shuts down the production of more cortisol. Unfortunately, lasting stress can also interfere with that feedback loop by damaging the neurons that turn off cortisol release, making future stress responses harder to stop.

Persistent overproduction of cortisol leads people to gain visceral fat, the type located inside the abdomen that is strongly linked to inflammation and health problems.[1] Such fat is prominent in so-called apple-shaped people, often men and postmenopausal women, as revealed by a high waist circumference. Visceral fat isn't always obvious, though, as some thin people also have a lot of it. This type of fat increases with age, especially in men. In contrast, people whose fat accumulates below the skin, typically on their hips and thighs (known as pear shaped), have a lower risk of lifestyle-related health problems. Fat under the skin produces two to three times more leptin per fat cell than visceral fat. Because leptin tells the brain's energy-balance system how much stored energy is available, that makes visceral fat less effective at instructing the brain to stop storing more fat.

Visceral fat cells have more receptors for cortisol than fat cells beneath the skin, which might explain why stress has a specific effect on visceral fat. One study compared pairs of identical twins, in which one was on average thirty-seven pounds heavier than the other. Only the pairs in which the heavier twin stored fat around the abdomen showed a difference in their biological indicators of stress.[2] Specifically, the heavier twin had higher cortisol and more disrupted sleep in those pairs, whereas no differences in stress signals were found between twins when the heavier twin was pear-shaped. Many people with visceral obesity show biological signs of long-term stress, including increased responsiveness to stressors, which can be a lasting consequence of stressful early life experiences.

In laboratory animals, long-term stress is a great way to cause metabolic syndrome.[3] Just putting stress hormones in the drinking water causes weight gain and increases blood sugar and cholesterol in mice.[4] Monkeys that are persistently stressed show weight gain around the abdomen, insulin resistance, changes in cholesterol, and high blood sugar. They are also more susceptible to heart disease. Rats that are stressed by housing with a socially dominant partner lose weight initially, but when the other rat is removed, they gain the weight back as visceral fat, which increases with every round of stress and recovery.[5]

Persistent stress also increases the risk of metabolic syndrome in people. One study found that higher cortisol at the beginning of the research predicted higher weight, higher blood pressure, and more insulin resistance five years later.[6] A thirty-five-year follow-up of middle-aged Swedish men found that persistent stress increased their risk of diabetes by 50 percent.[7] Another study followed sixteen sailors participating in an around-the-world race.[8] Over five stressful months at sea, they lost an average of ten pounds, and the weight they regained afterward was mainly visceral fat. Their blood chemistry after the race showed early signs of metabolic syndrome, and their blood pressure had increased slightly.

Lasting stress also activates the sympathetic nervous system, which orchestrates the fight-or-flight response to danger. A neurotransmitter called neuropeptide Y is released from this system when it is persistently active, in rodents and people. In mice that eat a high-fat diet, physical or social stress leads to visceral obesity within two weeks and metabolic syndrome in three months. All these consequences are greatly reduced by inhibiting the actions of neuropeptide Y on visceral fat, even as the mice continue to overeat.[9] Although this process hasn't been examined in people, it's plausible because we also have receptors for neuropeptide Y in our visceral fat that promote fat storage.

Taken together, this research indicates that stress contributes directly to many of the health problems that are often attributed to obesity, including diabetes, high blood pressure, high cholesterol, and heart disease. In theory, stress reduction could go a long way toward improving health outcomes. But in practice it's rarely that simple.

The Stress of Social Stigma

While we can influence some stressors, like a high-pressure job, others are out of our control. In particular, the stress resulting from other people's attitudes is hard to fight. It seems unfair to blame people for their weight while ignoring the possibility that bias and discrimination may cause some of the health problems that we attribute to obesity.[10] Stigma is an important cause of health inequalities, in part because it affects so many people worldwide. Until we stop discriminating against people with obesity, there is no way to tell how much of its health risks are due to the stress caused by social stigma.

One estimate suggests that the social environment (how other people treat us) is responsible for more than half of the individual variation in health outcomes, while lifestyle choices account for less than a quarter.[11] People with low social status are especially likely to get metabolic syndrome. The Whitehall studies, a long-running examination of factors predicting health in British civil servants, found a strong association.[12] Women who were not overweight initially but experienced three or more periods of workplace stress in fifteen years were five times more likely to have metabolic syndrome than those who were less stressed. Men who had at least three periods of stress were twice as likely as other men to have metabolic syndrome. Differences in health behaviors, like smoking or exercise, accounted for only about a third of the differences in health outcomes in the Whitehall studies.

A powerful source of stress in people is social subordination, being disrespected or dismissed based on our place in society.[13] Though our primate relatives also devote time and energy to keeping subordinates in their place, people are the champions at making others of their species miserable. Think of all the ways we can communicate that someone is beneath us, from active insults to ignoring their attempts to contribute. Widespread weight discrimination, along with everyday expression of negative stereotypes and hostility, add up to a strong experience of social stigma for many obese people. Shame, a common reaction to social stigma, reliably causes the release of stress hormones.[14] Obese adults have almost four times as much cortisol in their hair, where it accumulates due to long-term stress exposure, as overweight or normal-weight people.[15] By ages eight to twelve, obese children already have more cortisol in their hair than normal-weight children.[16]

Many women feel stressed by exposure to antifat bias even if it isn't directed at them. In one study, women who saw a collection of video clips from (real) television shows making fun of fat people showed elevated cortisol, whether or not they were overweight themselves.[17] This may happen because body dissatisfaction is so widespread among women of any weight.

A large amount of research suggests that discrimination leads to poor health.[18] In various groups, including African Americans and Asian Americans, perceived racial discrimination predicts an increased risk of heart disease, high blood pressure, and other illnesses, as well as increased odds of obesity. A potential biological link between discrimination and these disorders is long-term inflammation, which is increased in people who experience discrimination and contributes to the disorders caused by discrimination. Weight discrimination is associated with higher blood levels of C-reactive protein, a marker of inflammation that predicts diabetes and cardiovascular disease, in normal- and overweight people.[19] The relationship was not seen in severely obese people, perhaps because they have high baseline levels of C-reactive protein. In another study, obese women who experienced more weight stigma had higher cortisol and more oxidative stress, which causes inflammation, even after the researchers took into account the participants' weights.[20]

Persistent stress is one reason that obesity is more common among poor people. As early as age ten, children from families with the least money and social status have cortisol levels that are twice as high as those of children from the most prestigious families.[21] Discrimination also reduces willpower, because it takes a lot of mental effort to suppress the urge to react negatively to poor treatment. In laboratory studies, middle-class people who are treated as if they're not important find it more difficult to concentrate, ignore distractions, and resist inappropriate behavior. Because their time and money are so tight, poor people have to make many important decisions, while being disrespected at every turn. As a result, they have less willpower left over for weight control, which most of us find stressful enough already.[22]

That experience sounds awfully depressing, and indeed stress, depression, and obesity are tightly linked. Long-term studies suggest that the relationship goes both ways, with depression and obesity each making the other about 50 percent more likely over time.[23] Depression leads to inflammation and persistent production of cortisol.[24] Body dissatisfaction is a risk factor for both depression and obesity. Given those similarities, it seems unsurprising that they tend to occur in the same people. Depression also predicts many of the same diseases as obesity, including heart disease and diabetes.

Stress and Overeating

The stress of dieting or body dissatisfaction often causes weight gain and thus more stress. People who are underweight or at the low end of the normal range typically lose weight when stressed, while those of us at the high end of the normal range or above are more likely to gain.[25] Given that the brain doesn't care why a person is starving—a famine and a diet feel the same to the energy-balance system—we shouldn't be surprised that dieting increases the release of stress hormones.[26] Stress leads to weight gain because stress hormones promote eating by increasing hunger and reducing the effectiveness of cues that signal fullness.

This pattern is particularly pronounced in controlled eaters. People who try to carefully monitor their food intake have higher levels of stress hormones than intuitive eaters (those who eat when they're hungry and stop when they're full), even if the controlled eaters are not dieting when the hormones are measured.[27,28] People who score higher on a test of controlled eating also have shorter telomeres on their chromosomes, a consequence of cellular aging that can be accelerated by lasting stress or various metabolic disorders.[29] This damage appears to be preventable, however, as people who get regular exercise don't show any effect of lasting stress on telomere length.[30]

Stress makes food more rewarding. Activation of the stress-hormone system leads to the sustained release of endogenous opioids, neurotransmitters of the brain's pain-relief system, which act on the same receptors as opiate drugs like heroin. The release of endogenous opioids amplifies the rewarding power of tasty food, so that the sequence "stress leads to comfort food, which leads to feeling better" is likely to be remembered and repeated. Perhaps in a similar way, stress also increases the risk of addiction or its relapse.[31] That's the stress trap: dieting makes people stressed, which makes it easier to gain weight, especially after the diet ends.

Part of the link between stress and overeating may be the signaling molecule ghrelin, which is produced in the stomach and acts on the brain to promote feeding.[32] In rodents and people, stress increases the release of ghrelin, which contributes to comfort eating by increasing the reward value of calorie-rich food. Stress does not lead to comfort eating in mice lacking the gene for ghrelin. This signal may be responsible for the junk-food cravings that are common at times of stress. People eat more calorie-dense, tasty foods, especially snack foods, after they are given stress hormones in the laboratory, and so do people who have a strong response to stressful experiences. In one study of thin young men, injections of stress hormones increased their calorie intake substantially.[33] Comfort eating does make people and other animals feel better, at least temporarily, by lowering stress hormone release.[34]

In rodents, scientists understand some of the brain changes that link dieting to stress-induced binges, making that cycle self-reinforcing.[35] Three weeks of food restriction in mice, resulting in a 10 to 15 percent weight loss, leads to an increase in stress hormones and stress-related behavior. In brain regions involved in feeding and stress, weight loss causes gene changes that increase the ability of mice to produce stress hormones, making them respond more strongly and persistently to future stressors. A key point is that these changes are not reversed when the mice are again allowed to eat freely and their weight returns to normal.

Similar effects were found in mice that were switched from a high-fat to a low-fat diet without restricting their caloric intake. Eight weeks later, the mice that had dieted remained more likely to binge on high-fat foods when exposed to stress. High-fat foods caused them to produce hormones that increase hunger, especially reward-driven eating, which did not happen in mice that hadn't dieted. So losing weight caused changes in the brains of mice that led them to become more susceptible to stress and more likely to binge on high-fat foods in response to it. That response is the opposite of what a typical dieter would want, but it would be a survival-promoting adaptation to an environment where famine is common: "Hard times are coming, and food is available, so I'd better stock up now."

Sleep Loss Makes Us Gain Weight

Another way that stress leads to overeating is by causing insomnia, which in turn creates more stress. Cortisol increases in proportion to sleep deprivation, consistent with the idea that it is stressful. The sympathetic nervous system, which controls the fight-or-flight response, is also more active in sleep-deprived people. Sleep loss leads to weight gain by giving people more waking hours in which to eat and by fooling the brain's energy-balance system into thinking that the body is losing weight.

Most of us don't get enough sleep these days.[36] People who are forced to spend ten hours in bed sleep for about nine hours, which researchers consider to be our physiological sleep need. In 1960, the typical sleep duration for adults was similar, at eight or nine hours. By 1995, it had dropped to seven hours. In 2004, 30 percent of adults between thirty and sixty-four years old reported sleeping less than six hours a night. When sleep is tracked with a wristband, the average sleep length since 1990 is even shorter, around six hours. This change in sleep habits overlapped with the increase in obesity and may have contributed to it.

Lack of sleep predicts later weight gain in children and young adults through their twenties.[37] Seven long-term studies followed thousands of children, starting between birth and age nine.[38] All the studies found that the less sleep the children got, the more likely they were to become overweight between three and twenty-seven years later (depending on the study). In middle-aged and older adults, sleep loss is not strongly linked to future weight gain, perhaps because most adults have already gained as much weight as their sleep duration encourages. Extremely short sleep durations, less than five hours, may increase weight in older adults. Shift work interferes with sleep cycles and causes cortisol secretion, which can lead to weight gain. The long-term cortisol of shift workers correlates with their weight.[39]

Short sleep affects several hormones that increase hunger. Sleep deprivation reduces leptin in the blood, leading the brain's energy-balance system to think that people have lost weight and take the usual steps to make them gain it back.[40] In one laboratory study, healthy young men were allowed only four hours of sleep every night for six days, followed by a week of twelve-hour nights for recovery. After sleep deprivation, their blood levels of leptin were 20 percent lower than after recovery.

Another way that sleep deprivation increases appetite is by causing the release of ghrelin. In laboratory and large population studies, people who get four to six hours of sleep a night show lower leptin and higher ghrelin levels in their blood than those who get eight to ten hours. Probably as a consequence, they feel hungrier and snack more in the evening. Sleep deprivation also impairs willpower, making it more difficult to resist this increased hunger.

In an example of how the correlation between obesity and poor health can result from hidden factors that cause them both, sleep loss also increases the risk of developing type 2 diabetes, independent of weight. That happens because insulin's ability to control blood sugar is impaired by sleep deprivation, within a week of five-hour nights. A meta-analysis combined data from ten studies that followed more than 90,000 people for at least three years.[41] Trouble staying asleep when the study began predicted an 84 percent increase in the risk of being diagnosed with diabetes before it ended. Other risk factors included trouble falling asleep (48 percent increase) or sleeping less than six hours a night (28 percent increase). This association between sleep loss and diabetes was confirmed by a huge study that followed almost 175,000 people (ages fifty to seventy-one when the study began) over three to ten years.[42] Sleeping less than five hours per night at the study's start increased the risk of becoming diabetic before its end by 46 percent. This relationship did not depend on weight gain, as it occurred in people of all weights. Daytime naps longer than an hour also increased diabetes risk by 55 percent, regardless of nighttime sleep length.

Stress and sleep loss strongly promote weight gain and the types of poor health that are associated with obesity. Although a clinical trial is currently investigating whether increasing sleep duration might lead to weight loss in adults, that outcome seems unlikely considering how long-term weight gain increases the defended range of the brain's energy-balance system. Far more likely is that getting enough sleep early in life may reduce the odds of weight gain and health problems later on.

How to Manage Stress

Avoiding stress is a tricky proposition. Experience with moderate, controllable stress is helpful because it increases resilience, so we don't want to keep away from all stress. Plus we've been told so many times already that stress causes all sorts of health problems that for many people the advice itself probably leads to cortisol release at this point. But there are some research-tested ways to relax under trying circumstances, including meditation and exercise—both of which are good for health in many other ways as well.

A stress-reduction program created for people living with chronic pain is widely available and effective for many types of stress. Designed by Jon Kabat-Zinn of the University of Massachusetts, the Mindfulness-Based Stress Reduction program teaches mindfulness through meditation and yoga over eight to ten weekly sessions and one full-day retreat, with daily meditation practice at home in between. Mindfulness helps people become less reactive as they learn to avoid borrowing trouble from the past or future, instead focusing on the present moment without wishing for it to be different. This course focuses on helping people learn to tend and befriend their bodies, as they are right now, without judgment. For anyone who feels overwhelmed by persistent stress, learning how to handle it gently can be a big help. Many studies show that this course builds people's ability to cope with distress and disability, producing medium to large improvements in anxiety, depression, and quality of life for most participants.[43] For anyone who can't find a class or afford one, the course is also available as a book by Kabat-Zinn, *Full Catastrophe Living: Using the Wisdom of Your Body and Mind to Face Stress, Pain, and Illness.* He has also recorded guided meditations from the course.

Whether it is possible to reduce exposure to stress or not, lifestyle changes can help break the inflammatory cycle that leads to disease. A few studies support the idea that eating omega-3 fatty acids, anti-inflammatory components of foods like salmon or walnuts, can help to prevent inflammation from leading to metabolic syndrome. Eating more omega-3 fats is unlikely to hurt, in any case. Exercise also reduces inflammation,[44] which may explain why it provides extra benefits to obese people and elderly people, as both groups have more inflammation than average.[45] That may be one reason that people tend to gain weight as they get older. In mice, at least, exercise protects against inflammation in the hypothalamus resulting from a high-fat diet.[46] In addition to decreasing visceral fat, exercise triggers muscles to produce anti-inflammatory compounds. For stressed bodies, exercise really is the best medicine.

8

When Calories Don't Count

When nutrition professor Susan Roberts was designing a diet program, she tested the recommendations on herself. If she could maintain her normal weight on the meal plans, then she felt confident that they would work for her readers as well. The plans worked as intended, but only as long as she cooked at home. When she ate purchased entrees that supposedly contained the same number of calories, her weight started creeping up. To figure out why, she measured the energy in various low-calorie frozen dinners, which turned out to average 8 percent more calories than the labels claimed.[1] Those averages are within the limits set by the FDA, which requires that grocery calorie labels be accurate only to within 20 percent.

This inaccuracy is a problem if weight loss is just math, as we often hear. The advice is all over magazines and the Internet: cut 100 calories a day to lose 10 pounds a year. The logic is that 3,500 calories is the amount of energy in one pound of fat, so 100 calories times 365 days equals 10.4 pounds. There are a bunch of problems with this reasoning, which relies on a naive view of weight regulation. To name a few, once people get below their defended range, their bodies will burn less energy due to metabolic suppression, enough to overwhelm the missing hundred calories. In addition, as people lose weight, fewer calories are required to support their lighter bodies, so the calorie deficit goes to zero over time. A dynamic model of how energy imbalance affects the body (without taking metabolic suppression into account) suggests that it would take more like three years, on average, to lose 10 pounds after cutting 100 calories—and the pounds will stay off only if the calorie reduction is also permanent. In any case, most people probably don't care about the average, but about what will happen to them personally. Individual variations in physiology, based in part on each person's population of gut bacteria, lead to huge differences in weight loss from the same calorie deficit. Finally, almost no one can count accurately enough to figure out how to cut 100 calories, let alone maintain that accuracy for years. The idea of counting calories appeals to our inner nerd because it feels rational and quantifiable, but getting accurate numbers turns out to be much harder than it looks.

Not Just Math

The information on grocery store labels looks precise—152 calories per six-ounce serving sounds like an exact measurement—but provides only an approximation of how much energy a particular person gets from eating that food. The reasons are found at every step of the process, from measuring calories to cooking food to digesting it. Considering that a 5 percent error in energy balance over time would be enough to cause noticeable weight gain, such uncertainty is an important consideration in deciding whether to manage food intake by counting calories. At the minimum, people would have to figure out their own energy needs by experimentation, based on the characteristics of their own bodies, which foods they habitually eat, and how they prepare those foods.

Obviously no one can deny the first law of thermodynamics: energy entering the body has to go somewhere. It can't just disappear. But most "calories in/calories out" arguments about weight loss ignore vast individual differences in how bodies handle energy. They make as much sense as saying, "Look, a Humvee is basically a Prius. Just put gas in them, and they go." The people buying the gas might be inclined to disagree. Two people's response to the same energy deficit can be just as different.

Most people realize that food energy can fuel physical activity or be stored as fat, but the possibilities don't end there.[2] Food can also be excreted from the body because it has failed to be digested or absorbed. Energy can be used to grow or repair body tissues, including in pregnancy and breast-feeding. A lot of it goes to basal metabolism, the cost of keeping a body idling when it's not doing work. That energy expenditure depends on body size and the proportion of muscle versus fat, along with an individual variation of about 250 calories per day among people of the same size.[3] Contrary to popular belief, it is not affected by how many meals people eat or when they eat them. Finally, energy is wasted as heat during digestion, especially of protein, in what is called the thermal effect of food, which also differs across individuals. That reaction can be seen in the need to take off a sweater while eating a big meal. As weight is lost, all three components of energy expenditure—basal metabolism, the calories burned by a given amount of physical activity (because it costs more to move a larger body), and the thermal effect of food—decline until a new equilibrium is reached, at which point people stop losing weight unless they cut more calories.

The Trouble with Labels

The problems with calorie counting begin before the first bite. Food labels can be highly inaccurate, as filmmaker Casey Neistat discovered when he analyzed some of his favorite meals.[4] A tofu sandwich advertised as healthy turned out to have almost twice its labeled calories when it was tested in the laboratory. Food producers have an incentive to claim lower calorie counts to get health-conscious consumers to buy more of their products. When Susan Roberts and her research group went to chain restaurants and purchased items labeled as low in calories (under five hundred), the portions averaged 18 percent more calories than advertised, up to twice as many.[5] The free side dishes often had as many calories as the entree they accompanied, but because each item was listed separately, a customer would have to look carefully to notice that fact. The take-home message seems clear: a person could gain a lot of weight relying on calorie labels.

Getting calorie counts right is also hard for meals cooked at home. To determine how many calories a food contains, the most accurate method uses a bomb calorimeter. That's pretty much what it sounds like, a sealed chamber in which electricity burns the food to release its energy as heat, measured by a temperature increase in a surrounding water bath. The idea that a pound of fat contains 3,500 calories comes from bomb calorimetry measurements. To label every food in the grocery store, though, that procedure would be time-consuming and expensive.

Instead the label numbers come from a century-old system developed by Wilbur Atwater. He fed people diets with differing proportions of carbohydrates, fats, and proteins and used a bomb calorimeter to measure the calories in the food they ate and the waste they excreted. Then he calculated how much energy his subjects were getting from carbohydrates (four calories per gram), fats (nine), protein (four), and alcohol (seven). Aside from a few tweaks, for instance to account for high-fiber foods, scientists still use his formula today.

Not every gram of carbohydrate is created equal, however. Nor protein, nor fat. At every step of extracting the energy from foods, some is lost. How much? Well, it depends. Chemical reactions in the gut break food down into sugars, fatty acids, and amino acids that can be absorbed into the blood. The efficiency of these reactions depends on many factors, which may be specific to the food or the individual or both.

Break It Down Now

To digest food, enzymes in the gut must contact it. The smaller the pieces that end up in the stomach, the more thoroughly that food will be broken down. Chewing helps with that, and so does chopping. A pureed carrot yields more calories than a whole one, for instance. These differences can be large. Almonds have a calculated value of 170 calories, but people extract only 130 of those calories, probably because some of the fat is not available for digestion.[6] Similarly, people who eat peanut butter absorb more fat calories than those who eat whole roasted peanuts. The Atwater numbers in general are close to the values of processed foods, but can be substantially off for whole foods.

Cooking releases extra energy from food by breaking down cell walls and other structures that lock energy away from digestion.[7] That may explain why cooking is used in every human culture that's ever been studied. Richard Wrangham at Harvard believes that when our ancestors figured out how to cook, they got so much extra energy that they were able to support large brains for the first time in evolutionary history. For comparison, researchers calculate that if we ate the same diet of raw plants as other primates, people would need to eat for 9.3 hours per day just to get enough calories.[8] People who eat nothing but raw foods are indeed a lot thinner than vegetarians who cook.[9] Half the women who eat only raw food have such low body fat that they don't menstruate, suggesting that their bodies aren't getting enough energy.

The effect of cooking is largest for starch. A raw potato, for example, has half as many usable calories as a cooked whole potato, and a handful more calories can be extracted from mashed potatoes. Cooking wheat makes a third more calories available. Raw meat also has fewer available calories than cooked meat. To estimate the difference, researchers measured oxygen consumption in Burmese pythons.[10] Cooking reduced their cost of digestion by 12.7 percent, grinding the meat reduced it by 12.4 percent, and the combination by 23.4 percent. In people, cooking egg whites, compared with eating them raw, increases the amount of digestible protein by 50 percent.[11] Similarly, eating peanuts roasted instead of raw allows mice to harvest more of their fat content.[12]

Another reason we don't get all the energy out of our food is because digestion itself uses energy. Fat is the easiest macronutrient to digest, with losses of only 3 to 5 percent, meaning that when people eat a hundred calories of fat, they get about ninety-six usable calories. Carbohydrate digestion costs 5 to 15 percent of the food's energy, so that a hundred calories of food yields about ninety calories. Proteins are the hardest to digest, costing 20 to 35 percent of their energy, so a person might get as few as sixty-five calories from eating a hundred calories of protein. People who believe they're not cutting calories on their low-carb diet might just be mistaken, given how hard it is to digest meat.

If you're starving and need to extract every possible calorie from food, make sure it's low in fiber and as processed as possible. In long-term studies, weight gain is best predicted by eating French fries, potato chips, and sugared soda.[13] The time and energy saved on food preparation may be lost later on to illness, though. On the other hand, to get the maximum possible health benefits and feelings of fullness from eating, the evidence suggests concentrating on whole foods that are cooked at home. In the same long-term studies, avoiding weight gain is predicted by eating whole grains, vegetables, fruits, and nuts.

Your Bacterial Passengers

The last factor that contributes to individual variability in how many calories a person extracts from food is the gut bacteria population. The human gut contains tens of trillions of bacteria; many more of our cells are bacterial than human. We support these hitchhikers because, like other animals, we need gut bacteria to digest our own food. These resident bugs come from three main groups, called Bacteroidetes, Firmicutes, and Actinobacteria, which break down parts of our food that would otherwise be wasted. The Bacteroidetes even synthesize vitamin K, a necessary nutrient that we can't make for ourselves. These bacteria don't just nibble around the edges of our diets, either. Their contribution is large. Without them, we would have a much harder time meeting our nutritional needs. Gut bacteria are also important for the normal development of the immune system, as they influence the initial population of T cells, the white blood cells that respond to cellular damage or infections.

Researchers can produce "germ-free mice" without any gut bacteria by birthing them through Cesarean section (newborns get a lot of their initial bacteria colony from the mother's vagina) and raising them in an isolation chamber. Germ-free mice eat 29 percent more food but have 42 percent less body fat than similar mice that were raised normally.[14] When fed a tasty high-fat, high-carbohydrate diet, the germ-free mice eat as much as normal mice but gain much less weight. Eating more and weighing less might be a dieter's dream, but it would have been a serious disadvantage when food was hard to get, throughout most of human history. Bacteria release a lot more energy from food than our own digestive enzymes can extract. If germ-free mice receive transplanted gut bacteria from a normal mouse (yes, that's a polite way of saying a poop transplant), they start eating 27 percent less food, the same as a normal mouse, while increasing their body fat by 60 percent in two weeks.

Although not having any bacteria at all is an extreme experimental manipulation, the gut microbiome is a vibrant living community that can be transformed through food choice, illness, or antibiotic use. For example, researchers put people on a diet of meat, eggs, and cheese, giving them 70 percent of their calories from fat and the rest from protein, with almost no fiber.[15] Within a day, they had more Bacteroidetes, which are tolerant of bile, a secretion involved in fat digestion. At the same time, the Firmicutes, which digest plant starches, went down. A whole-grain, vegetarian diet had the opposite effects. These changes are similar to differences between meat-eating and plant-eating species. When the participants resumed their regular eating habits, their gut bacteria returned to normal in two days.

Bacteria Influence Weight

Gut bacteria have wide-ranging effects on energy pathways throughout the body, and the relative abundance of different species varies with body weight. Mice that are obese because they cannot make leptin have 50 percent fewer Bacteroidetes and 50 percent more Firmicutes compared with their thin counterparts. A similar pattern is seen in mice that became obese on a high-fat/high-sugar diet, as well as in obese people. When germ-free mice are transplanted with gut bacteria from an obese mouse or an obese person, they gain about twice as much weight as when the bacteria come from a lean mouse. These differences are reversed when animals lose weight on either low-fat or low-carbohydrate diets, at least while the weight loss lasts.

The relative prevalence of gut bacteria can also change over generations, across the population. The species *Helicobacter pylori* was the most common resident of the stomach since ancient times, making up more than half of all stomach bacteria, but it has steadily declined in people in developed countries around the world over the past century.[16] In 1900, more than 80 percent of people carried this species, but by 2000 only 6 percent of children in the United States, Germany, and Sweden had it. The antibiotics often used to treat childhood ear infections wipe out this species in 20 to 50 percent of people who receive them, which explains some of the decline. Because *H. pylori* produces a lot of acid, its loss is a mixed blessing, increasing esophageal reflux and associated disorders but decreasing stomach cancers and ulcers. The disappearance of this species also changes the production of hormones involved in regulating appetite, reducing leptin, at least in the stomach, and increasing ghrelin in the blood, which may help to set the brain's defended range in early life. Successful eradication of *H. pylori* by antibiotic treatment in adults is often associated with weight gain, relative to cases in which the same antibiotics fail to kill their intended target.[17]

Some types of gut bacteria are influenced by their host person's genes, while others are not. The prevalence of bacteria from the Bacteroidetes family responds to diet and does not depend on the person's genetic heritage. In contrast, a twin study found the highest heritability in a family called Christensenellaceae, which is not easily modified by diet.[18] When transplanted to germ-free mice, a member of this family caused weight loss and reduced body fat. In another study, germ-free mice were given gut bacteria from human twin pairs with one lean and one obese twin.[19] When the mice were housed separately, they became lean or obese, just like the twin whose bacteria they had received. When the pair was housed together, the lean twin's bacteria became established in both mice, and the obese mouse lost weight, but only on a diet low in fat and high in vegetables. If the diet was high in fat and low in vegetables, the obese mouse stayed obese.

The bacteria types common in obesity lead to weight gain because they help us get more energy from food.[20] They are better at breaking down complex carbohydrates like cellulose, xylan, and pectin. For people who eat a lot of fruits and vegetables, that improvement in digestion may provide an extra 140 to 180 calories per day of usable energy. Gut bacteria influence the absorption of nutrients in a variety of ways, for instance by slowing the movement of food through the gut to allow more complete extraction of nutrients, and by increasing the production of an enzyme that moves glucose from the small intestine into the blood. They also suppress an enzyme called lipoprotein lipase that normally limits the ability of fat cells to take up fatty acids and triglycerides from the blood, resulting in more fat storage. This mechanism seems to be particularly important for obesity, as germ-free mice gain only 10 percent of their weight instead of 60 percent if a regulator of this pathway is blocked after they receive the gut bacteria transplant. Gut bacteria also reduce the use of fat for energy in the liver and muscles.

Weight-loss surgery may take advantage of these pathways to change the defended range of the brain's energy-balance system. Gut bacteria modify signals of hunger and fullness, so they might act on the brain's energy-regulation system. Weight-loss surgery causes changes in gut bacteria in both mice and people, similar to those seen during diet-induced weight loss, and also reduces diabetes symptoms.[21] Both these effects of bariatric surgery occur before weight loss and are independent of diet. When bacteria from mice that have had surgery are transplanted into germ-free mice, they lose weight, even though they are very thin already.

In one person, a gut bacteria transplant may have caused obesity.[22] Such transplants are an effective way to cure infections by *Clostridium difficile*, which can take over when the other gut bacteria have been killed by antibiotics, causing almost nonstop diarrhea. After multiple attempts to eradicate her *C. difficile* infection with antibiotics failed, a thirty-two-year-old woman agreed to try a fecal transplant, which cured her infection. At that time, she weighed 136 pounds and had never been obese. Her daughter, the donor, weighed 140 pounds before the transplant, but her weight shot up to 170 pounds soon afterward. The mother also gained weight, despite repeated attempts to control it with diet and exercise, reaching 177 pounds by three years after the transplant. Given that both of them gained weight at the same time, their shared gut bacteria seems to be the most likely explanation.

Gut Bacteria for Weight Loss?

Although researchers understand some of the details of how gut bacteria influence weight, they have not yet figured out how to apply that knowledge to make people lose weight. One possibility is a bacterial species called *Akkermansia muciniphila*, which becomes much more prevalent after weight-loss surgery and prevents mice from gaining weight on a high-fat diet. A clinical trial of this species is under way in people, but it will be a while before microbiome therapy for obesity is ready for action.

Researchers are also testing a drug that mimics the effects of weight-loss surgery on gut bacteria, at least in mice. After surgery, human patients show a substantial increase in bile acids circulating in the blood. In addition to their main job of helping to digest fats, bile acids released during meals activate a receptor called FXR, which regulates fat and sugar metabolism. Both directly and through FXR, bile acids also influence the abundance and composition of gut microbes. In mice that are genetically modified to lack this receptor, weight-loss surgery causes only a transient decrease in weight, which is fully regained within five weeks. (This study provides more evidence that the effect of surgery is not due to limiting meal volume, as these animals were able to gain weight despite their surgically reduced stomach size.) Gut bacteria also react differently to surgery in mice lacking FXR.[23] They do not show the usual reduced abundance of Bacteroidetes or the increased abundance of a bacterial group called Roseburia, which is low in people with diabetes.

A drug called fexaramine activates the FXR receptor, mimicking the effects of bile acids. When mice swallow this drug, it acts in the intestine to reduce weight gain in response to a high-fat diet, without decreasing the amount that animals eat.[24] Instead the drug reduces inflammation and production of glucose by the liver. It also increases insulin sensitivity and basal metabolic rate, as the mice burn enough extra sugar and fat to raise their body temperature by 2.7 degrees Fahrenheit. These effects must result from activation of the FXR receptor, because they don't occur in drug-treated mice that lack the receptor.

A critical difference between animal research and clinical therapy is that doctors can't transplant bacteria into germ-free people. Instead researchers have to contend with the species that are already present in the gut, an assortment that is different for each individual. Changing the microbiome is less like giving a drug and more like trying to restore a damaged habitat, with a lot of potential interactions between species that are difficult to predict. For example, when scientists removed all the starfish from a stretch of shoreline, the diversity of species was cut in half as mussels took over much more territory than before, successfully crowding out their competitors. Different species of bacteria also compete with one another, so if one type is too prevalent, it might be able to elbow out any replacement that doctors put in. The number of possible outcomes is head-spinningly complicated, so it will probably take a long time to work out the details.

Another issue with applying this research in the clinic is defining what a desirable population of gut bacteria would look like. As a first guess, we might be tempted to replicate the ancestral microbiome by copying the gut bacteria of modern hunter-gatherers. There are certainly some major differences. Hunter-gatherers have a much more diverse set of species than ours, with estimates ranging from 50 percent more to twice as many. But there are also some problems with the idea of copying their microbiome. Gut microbes vary across different populations of hunter-gatherers and change seasonally as food sources shift, suggesting that there was probably no single ancestral state. More important, the gut bacteria of these populations are adapted to their own lifestyles and may not play well with ours. Bacteria that specialize in digesting fiber would have trouble finding anything to eat in the guts of a typical suburbanite.

In Sickness and in Health

The gut microbiome also contributes to disease. Because gut bacteria help calibrate the immune system early in life—particularly its ability to distinguish self ("that's me, so leave it alone") from nonself ("foreign invaders must die")—changing the microbiome in adulthood might lead the immune system to attack the intruders, potentially causing autoimmune diseases.

Diabetes is associated with shifts in the gut microbiome, regardless of body weight.[25] Unlike normal mice, germ-free mice do not develop insulin resistance on a high-fat, high-carbohydrate diet. After receiving a bacterial transplant from normal mice, however, they show a strong increase in blood sugar and insulin resistance. This result makes sense because gut bacteria can influence the uptake of glucose into the blood. One group of gut bacteria may help to cause diabetes. These are the bacteria that produce a signal called lipopolysaccharide, which triggers the production of many inflammatory molecules, especially in animals fed a high-fat diet. The resulting inflammation in fat cells and elsewhere seems to produce both type 2 diabetes and obesity. Similarly, when gut bacteria from lean people were transplanted into obese people with metabolic syndrome, their insulin sensitivity increased within six weeks.[26] At six months, though, the gut bacteria had returned to their normal state, in line with the idea that a bacterial ecosystem can push back against changes.

Researchers isolated a single species of bacteria that produces lipopolysaccharide from the gut of a very obese person. Initially, when he weighed 385 pounds, this bacterial species made up 35 percent of his microbiome, but six months later, after he had lost 110 pounds, it was undetectable. When germ-free mice were transplanted with this species alone, they became obese and diabetic, but only if they ate a high-fat diet.[27] In another experiment, feeding mice a high-fat diet for a month increased the population of bacteria that produce lipopolysaccharide, raising its blood levels by two- to threefold.[28] The mice gained weight and started showing symptoms of diabetes. The same things happened when lipopolysaccharide was infused into mice eating a normal diet. In contrast, blocking the receptor that receives signals from lipopolysaccharide protected mice from the effects of a high-fat diet. These experiments suggest that it might eventually be possible to prevent the development of diabetes even in people who are unable to exercise or eat well, though such a drug remains a long way off.

Gut bacteria are also modified by pregnancy, when increased fatness and reduced insulin sensitivity help the mother's body to provide the energy needed for fetal growth and breast-feeding. A study of ninety-one women in Finland found that pregnancy has a huge effect on gut bacteria.[29] Proteobacteria and Actinobacteria became more abundant in the majority of women, with no overall change in the abundance of Firmicutes or Bacteroidetes. Third-trimester bacteria contained fewer species than first-trimester bacteria, which are more similar to nonpregnant gut bacteria. By the third trimester and continuing to one month after birth, the mothers' bacterial colonies were more individually distinct than they had been early in pregnancy. Their babies also had a wide diversity of bacteria at six months of age but reached adult levels of similarity across individuals by four years of age. At this age, their gut bacteria were similar to the mother's as well. When germ-free mice were transplanted with bacteria pooled from five pregnant women, the third-trimester sample caused them to gain one third more weight than the first-trimester sample, even though they ate the same amount of food. The third-trimester sample also reduced glucose tolerance in the mice. These changes increase the risk of diabetes, so they may contribute to the development of gestational diabetes in some women.

The number of different bacterial species in the gut may also matter. People with a less diverse set of gut bacteria are more likely to be obese and to have metabolic syndrome.[30] In fact, researchers can predict whether people are fat or thin more effectively from their bacterial genome (90 percent accuracy) than from the same people's own genes (58 percent accuracy). Diversity can be reduced by antibiotic treatment, which not only kills bacteria but also favors certain species over others.

We can get a hint of which species are more likely to survive by recalling that industrial agriculture has used low doses of antibiotics in feed or water to promote weight gain in meat animals for decades. The strategy works in cows, sheep, pigs, chickens, and turkeys. The younger the animals are when treated, the more weight they gain. When researchers give mice low-dose antibiotics from birth, they end up with a higher proportion of Firmicutes and fewer Bacteroidetes (the pattern associated with obesity) and have more fat tissue than their littermates without antibiotics.[31] When mice receive antibiotics for four weeks in infancy, their gut bacteria population gradually goes back to normal, but then in adolescence they develop obesity and metabolic problems, including accumulation of fat in the abdomen and the liver. Starting antibiotics earlier in life, before birth, leads to more weight gain.[32] Putting antibiotic-treated mice onto a high-fat diet produces the strongest effects of all, especially in males, which are more susceptible to diet-induced obesity.

A similar relationship may occur in people. For example, among Canadian babies who did not receive antibiotics in their first year of life, 18.2 percent were overweight or obese at age twelve.[33] In contrast, 32.4 percent of children exposed to early antibiotics became overweight or obese. After adjustment for potential confounding factors like birth weight and the mother's weight, the relationship remained statistically significant in boys but not girls. Boys were more than five times as likely to become overweight at twelve if they had received antibiotics and almost three times as likely to have a lot of visceral fat. Similarly, among more than twenty-seven thousand Danish children born to normal-weight mothers, antibiotics in the first six months of life predicted a 54 percent higher chance of being overweight at seven years.[34] However, the opposite result was found for children of overweight mothers, who were half as likely to be overweight at seven if they had received antibiotics, suggesting that the effect of these drugs may vary depending on the child's initial gut bacteria population.

Curious about my own gut bacteria, I joined the American Gut project and sent off a sample. Researchers are crowd-funding a look at the human microbiome and how it's influenced by diet and lifestyle. Individuals can find out about their gut bacteria while contributing to science. The report showed that my ratio of Bacteroidetes to Firmicutes was high, as expected in someone who's not obese. However, my gut also contained a high proportion (about 18 percent) of Proteobacteria, which includes many common infectious microbes, like *E. coli* and salmonella. Many of them produce lipopolysaccharide, which is associated with inflammation and metabolic disease. My diet is probably not to blame, as I eat a lot of fruits and vegetables and get more fiber than an average person in the United States. My husband, who shares meals, a house, and a dog with me, has only 2 percent Proteobacteria in his gut.

One possible explanation is a series of antibiotic treatments that I received about ten years ago for a stubborn infection, because gut bacteria don't recover as well from repeated antibiotic treatments as they do from single treatments. So far, in my late forties, I have no allergies and no signs of metabolic syndrome, leaving the significance of this pattern unclear. As this story illustrates, we can measure more characteristics of gut bacteria than we can interpret.

For now, we can take a couple of lessons from this research. Parents should minimize antibiotic use in children, especially in the first year of life, because changes in gut bacteria at that age can have lasting consequences. The average child in the United States receives ten to twenty courses of antibiotics before age eighteen,[35] increasing the risks of asthma, allergies, and inflammatory bowel disease, in addition to obesity and diabetes.[36] Someday we may have drugs or probiotics based on this research, but they're not coming anytime soon. In the meantime, we should all think twice about blaming people for their weight. The composition of our gut bacteria profoundly affects our bodies, yet no one knows how to use that knowledge to change them.

9

Blame Your Ancestors

Anamarie Regino's parents could hear her screaming for her father as she rolled down the hospital hallway, but fear of the armed guards kept them from responding.[1] The state of New Mexico had decided that Ana's extreme obesity and failure to lose weight constituted child abuse. At age three, when an average child weighs 30 pounds, she weighed 120, down from a high of 130 earlier that year. By two years of age, she had weighed 70 pounds. Because her grandmother could no longer lift her, her father, Miguel Regino, quit his job to stay home with the toddler.

Her mother, Adela Martinez-Regino, started taking her daughter to doctors as an infant, trying to figure out why the baby would take ten or twelve bottles of milk and then cry for more. They considered the rare chromosomal disorder Prader-Willi syndrome, but the symptoms didn't fit. Unlike those patients, Ana was one and a half times as tall as a normal child and had other odd symptoms like premature bone growth, early tooth development, and thick, adult-like hair. An endocrinologist suspected a leptin receptor deficiency, which doctors later ruled out as well.

Unable to diagnose the child's problem, but suspicious because Ana gained more weight at home than in the hospital, her doctors reported the family to the Children, Youth and Families Department. The social worker, Lisa Perez, annoyed Adela by insisting on speaking to her in Spanish instead of English, her native language. In the statement Perez filed with the court, she spelled Ana's name wrong and recommended that the child go to foster care because her parents couldn't understand the seriousness of her condition, "due to language or cultural barriers."

Ana was removed on August 25, 2000, and returned home on November 10 after the family's lawyer negotiated a settlement agreement. For months, the family was required to host a series of social workers, who sat on the couch watching them care for Ana until the state returned legal custody to the parents in January 2001. The social workers left the family alone after that, but Ana's medical problems continued. At eleven, she developed diabetes, and by twelve, she had grown to five feet three inches and three hundred pounds.[2] Her condition has never been diagnosed, but her history leaves little doubt that she has a genetic mutation that prevents her brain's energy-balance system from working correctly. Ana's case is an extreme example of a general principle: that some people gain weight more easily than others, largely because of their genetic heritage.

Why Cavemen Didn't Get Fat

The inborn differences among people become obvious as food supplies increase. Because famines outnumbered fast food in human evolutionary history, the brain's energy-balance system is optimized to deal with scarcity. Over the 200,000 years that anatomically modern humans have existed, food has been constantly available to the majority of people for barely a century—and even now, more than a billion people subsist on less than a dollar per day. For almost all of humanity's existence, most people suffered occasional hunger from crop failures, long winters, hunting parties that came home empty-handed, or seasonal variations in wild plants. People whose bodies couldn't handle a few weeks or months on limited calories didn't live long enough to become anyone's ancestor. Now most of us are facing the opposite problem. We get plenty of energy from our food, but its nutritional value has plummeted as technology has made eating smoother and easier, from agriculture to refined grains to manufactured foodlike substances. Over the same period, mechanization has reduced the need for energy expenditure in daily life, both at work and at home.

"If we had a time machine, it would be the world's best weight-loss program," says Yoni Freedhoff, a physician specializing in obesity at the University of Ottawa. When our ancestors were nomadic hunter-gatherers, moving as food availability shifted with the seasons, people faced serious barriers to overeating, which made it hard to gain a lot of weight. To get enough to eat, they had to know how to find edible plants and how to track and kill animals, both complicated and place-specific skills that take a long time to learn. Acquiring food often required traveling long distances, which took time away from other priorities like making clothing to keep warm. Not that our ancestors' lives were, as Thomas Hobbes put it, "solitary, poor, nasty, brutish, and short." Indeed, modern hunter-gatherers in good territories may work as little as three hours a day for their food, leaving plenty of time for socializing, child care, and making tools or art.[3] Still, the ease of getting dinner as a well-off hunter-gatherer is on a different scale from stopping at the gas station to buy a jumbo bag of Doritos.

This ancestral arrangement led evolution to select for brains that motivate people to acquire as many calories as they can and to limit physical activity where possible. Indeed, all sorts of animals, from crabs to crows, show optimal foraging patterns that allow them to get the most energy gain from the smallest energy expenditure. Imagine a person who had plenty of groceries but stored them all at a friend's house a mile away. If he had to walk to get there, that arrangement wouldn't keep him from eating when he was hungry, but it probably would cut down on mindless snacking in front of the television, which is how a lot of people gain weight. For many of us now, there's little or no cost to eating whenever it occurs to us, which is not a situation that our brains are prepared to handle. From that perspective, it's not surprising that some of us have gained weight; the question is why some of us haven't.

Despite the claims of Paleo diet enthusiasts, our ancestors ate a wide variety of foods, including carbohydrates. Starch is found on the teeth of fossil humans as far back as thirty thousand years ago, though the cultivation of grains is only ten thousand years old.[4] Hunter-gatherers range from nearly vegetarian to nearly carnivorous, depending on what foods are available, and they eat huge amounts of honey when they can find it.[5] Our ancestors were probably at least that flexible, considering that they managed to survive under a variety of conditions all over the world. Most plants and almost all animals that people ate in the Stone Age are extinct, and our recent ancestors have modified their replacements through selective breeding. Agriculture increased the size of starches and grains, and the meat in grocery stores contains a lot more fat than wild meats do.

Human genes have also changed in the last ten thousand years.[6] Researchers estimate that between several hundred and two thousand human genes (as much as 10 percent of the total) show signs of recent rapid evolution. For example, when people first learned to domesticate cows, eight to nine thousand years ago in Egypt, only children produced the lactase gene, so adults could not digest milk. As herding spread, so did two different genetic mutations that independently led to adult lactose tolerance in European and East African populations. Lactose intolerance remains common in people of Asian and West African descent, whose ancestors did not herd cattle. Animal domestication also led to the evolution of new immune-system defenses, as people first came into close contact with animals and their germs.

Genes related to digestion of various types of food, including starches and alcohol, have been recently shaped by dietary practices. The invention of cooking was correlated with changes in digestion, bitter taste receptors, tooth enamel, and jaw musculature. In the past century, though, our genes have changed much less than our food environment.

Weight Runs in Families

William Gibson famously said, "The future is already here. It's just not very evenly distributed." The same can be said of the recent increase in obesity, which is concentrated among people with a family predisposition toward weight gain.[7] Between 1975 and 2005, the weight of the average person in the United States increased by twenty pounds.[8] (Since then, adults in most countries, including the United States, have not gained any additional weight, though the media has been slow to share that news.)[9] That doesn't mean that each of us weighs twenty pounds more than our parents. For the most part, people whose parents were slender are slender themselves. Those of us from average-weight families might be five pounds heavier than our parents. People whose parents were overweight have added ten pounds or so to their inheritance, often enough to just barely edge them into the obese category. That explains how a twenty-pound average weight gain can change the country's population from 14 percent obese to 35 percent obese. People with obese ancestors, on the other hand, might be fifty or a hundred pounds heavier than previous generations. This pattern of weight gain suggests that the modern environment affects people differently depending on their genetic tendencies, an idea that is supported by a lot of research.

Weight is as heritable as height, though it's easier to see the genetic influences on height because it doesn't change much in adulthood. Most identical twins have similar weights and body shapes, even if they were raised in different families.[10] Children are more than ten times more likely to be obese by age seven if both their parents are obese.[11] The siblings of an obese person are three to seven times more likely to be obese than a randomly chosen person. Across individuals, 50 to 70 percent of the variation in body weight (relative to height) is attributable to their genes. Body-fat distribution, which affects body shape, is even more heritable, at 70 to 85 percent. Another twin study found the heritability of both weight and waist circumference to be 77 percent when food is abundant.[12] In addition, susceptibility to type 2 diabetes is heritable, with a two- to sixfold higher risk if a parent or sibling has the disorder.[13]

In our ancestral environment, genes probably mattered less to individual differences in weight because food availability was the limiting factor. Few people had enough food to find out whether their genes predisposed them to putting on weight. Under those conditions, no selective pressure would lead evolution to remove those genes from the population, because they would not affect survival. Indeed, there isn't much pressure to get rid of them now, as diseases like diabetes and heart attacks mostly affect people who have already raised their children. On the other hand, our history provided a lot of pressure to weed out genes that made it easy to lose weight during food shortages, a common and sometimes fatal condition.

In a few people, damage to a particular gene directly affects weight by preventing the production of a protein that is required for the energy-balance system to work correctly. But more commonly genes affect the development of the body, especially the brain, changing the control of weight in adulthood. Genetic influences on weight are not independent of the environment, or of behavior. Instead our genes shape our physiology, where we choose to live, the way we eat, and how much we move. Taken together, these characteristics explain the wide variation in susceptibility to weight gain in the modern environment.

For example, genes influence the function of the thyroid, which produces hormones that regulate metabolism.[14] Impaired thyroid function occurs in 1 to 2 percent of the population in areas with adequate iodine intake.[15] It's ten times more common in women than in men, and the risk increases with age. Low thyroid hormone, found in 8 percent of women and 3 percent of men, makes people susceptible to weight gain. Even variations in thyroid hormone that fall within the normal range are associated with weight differences among individuals, presumably because they cause variability in basal metabolism.[16]

Identifying the specific genes that influence human weight has proven difficult. Only a small percentage of obesity is due to heritable defects in single genes involved in weight-regulation pathways. In most people with such defects, the brain's energy-balance system lacks a critical part, leading to extreme obesity that begins in early childhood, the pattern seen in Anamarie Regino. For those people, eating in response to hunger does not stabilize body weight because the mutations interfere with the satiety signals that terminate meals. All such mutations have the same effects in mice and people, suggesting that what we've learned about obesity in mice is likely to apply to us.

In the mid-1990s, two Pakistani cousins arrived at a hospital in London for an obesity consult. The eight-year-old girl weighed 189 pounds, and the two-year-old boy weighed 62 pounds. Both had been normal weight at birth, but they badgered their parents for food, eating far more than their siblings. The doctors determined that both children had a genetic mutation that prevented the production of leptin.[17] Because the brain's energy-balance system wasn't receiving any information about the amount of fat stored in their bodies, it thought the children were dangerously thin and kept pushing them to gain more weight.

These children were lucky: they received an accurate diagnosis that suggested a cure. At age nine, the girl was treated with injections of leptin.[18] Her weight dropped from 208 pounds to 180 pounds over the next year, as she continued to grow taller, and she stopped feeling hungry all the time. Providing the missing signal to her brain fixed the problem. Other children with unrestrained appetites, likely due to their genes, have been subjected to weight-loss surgery as young as two and a half.[19] Obesity caused by a single gene mutation is rare—doctors know of a couple dozen people who lack a functional leptin gene worldwide—but shows how the normal human brain regulates weight.

Another single-gene cause of obesity removes the neurotransmitter used by the eating-inhibition neurons in the arcuate nucleus of the hypothalamus. Because another protein encoded by the same gene influences hair color, people with this type of genetic obesity often have red hair.[20] The receptor that receives signals from this neurotransmitter is the most common single-gene cause of obesity, accounting for 5.4 percent of severe childhood cases, which are more likely than adult cases to be due to single-gene mutations.[21] Overall, one adult per thousand has a mutation in this receptor gene.[22]

Another gene, which encodes a signaling molecule on the energy-sensing pathway, has effects that are a dieter's nightmare.[23] When it's damaged, people become hungrier, especially as children, burn less energy via their basal metabolism, and show severe insulin resistance. Mice that have one normal and one abnormal copy of this gene become obese only on a high-fat diet. A few other mutations that cause obesity in mice have also been identified in people, including mutations that prevent the production of a functional receptor for leptin, but such obesity is rare.

For most people, a large set of genes with small individual effects contributes to body weight. For example, someone with two copies of the higher-weight allele of a gene called FTO is on average five pounds heavier than someone with two copies of the lower-weight allele, starting around age seven.[24] (An allele is a variant of a particular gene. People have two alleles of each gene, one inherited from the mother and one from the father.) The biochemical action of this mutation is strange. Genes contain alternating regions called exons, which encode parts of a protein (to be connected later to form a complete protein), and introns, which do not encode proteins. The FTO mutation occurs within an intron, and its importance to weight seems to be due to its ability to control expression of another gene that is important to early brain development.[25] Mice lacking this second gene in their brains grow up to have a high basal metabolic rate and weigh 25 percent less than their littermates. Among the genes that cause individual variation in obesity, FTO has the strongest known effect, so it's no surprise that the weaker genes have proven hard to find.

Another type of genetic variation contributing to obesity is individual differences in the genes that code for uncoupling proteins. These proteins sit in the membrane of mitochondria and control whether they produce energy in a form that fuels cellular biochemistry or simply release it as heat. By doing so, they regulate body temperature and the thermal effect of food, the temporary increase in metabolism caused by eating. People with certain alleles of the genes that encode these proteins weigh more, on average, than people with other alleles, though again the effects are small.[26]

All together, the dozens of small-effect genes that have been identified account for only 2 to 4 percent of the variation in obesity in the population. One possibility is that a lot more such genes might await discovery. Another is that obesity-related genes may undergo new mutations during the person's lifetime rather than being inherited from the parents. Researchers need to examine a lot more people to figure out which of these explanations is correct. In the meantime, if you want to know your own genetic odds of obesity, look around at your next family reunion.

Except for patients with a damaged leptin gene, progress in understanding the genetics of obesity has not yet led to effective treatments, but researchers remain hopeful. When I mentioned Yoni Freedhoff's time-machine line to Rudy Leibel on the phone, he retorted, "Most of the people who want to lose weight wouldn't take that trade-off." They might be wiser to go forward in time, he suggests, "to when we figure out how to do this with a pill, which is going to happen, though I don't know when."

How Nature and Nurture Interact

By now some readers may be thinking that familial effects on body weight might be due to the environment. That is, children could end up the same weight as their parents because they picked up their parents' eating and exercise habits early in life. People often imagine that either nature or nurture must be responsible for a trait, but that intuition is wrong. To see how genes and environment can both be critical to a developmental process at the same time, let's think about an easier example: nearsightedness. Children of nearsighted parents are usually nearsighted too, as the heritability of this problem is as high as 90 percent in some twin studies.

But that's true only if the parents and children grow up in a similar environment. When people adopt a new lifestyle, the outcomes change. In 1974, a community of Inuit in the Canadian Arctic Circle had recently started formal schooling.[27] Among adults over thirty, only 4.5 percent were nearsighted. In contrast, 45 percent of younger adults were nearsighted—a tenfold increase. The change that struck the Inuit in one generation has since spread to many other countries.[28]

The rate in the United States went from 25 percent in the early 1970s to 42 percent three decades later. In a few societies, nearsightedness rates are now about as high as they can get. Among modern Chinese students in Singapore, 90 to 95 percent are nearsighted. These changes occurred much faster than genes in the population could change, so some new environmental factor must be responsible.

Research suggests that the culprit is too much time spent indoors. One study compared six- and seven-year-old children of Chinese ethnicity living in Sydney, Australia, against those living in Singapore.[29] The rate of nearsightedness was more than eight times lower in Sydney (3.3 percent) than in Singapore (29.1 percent), despite similar rates of parental nearsightedness. Children in Sydney spent 13.75 hours per week outside, on average, compared with 3.05 hours per week for children in Singapore. Similarly, children in the United States who spend 2 hours per day outdoors are four times less likely to become nearsighted than those who spend less than 1 hour per day. It does not seem to matter exactly what children do while they are outside, as the risk scales with time outside regardless of activity.[30] Earlier research blamed nearsightedness on "near work," such as reading or sewing, but among children who spend the same amount of time outside, the amount of near work makes no difference. Instead the weak association between near work and nearsightedness arises because time spent with a book tends to trade off against time spent outside.

Studies in laboratory animals indicate that outdoor time matters because bright light is needed for correct eye growth in young animals.[31] Why should eye growth depend on bright light? There's no reason that it should, but for most of human history that caused no problems because indoor environments didn't exist. By inventing them, we outsmarted our own evolution, which had cobbled together a less-than-perfect solution that worked reliably enough, based on the circumstances at the time. Evolution does a lot of that, and people excel at changing their environment enough to reveal such shortcuts in our biology.

Despite this huge environmental influence, genes still affect nearsightedness. Children with a genetic predisposition are more likely to become nearsighted—or severely nearsighted—than children whose parents have good vision.[32] Here's one way to think about it: picture a hundred children lining up to receive their portion of nearsightedness. In an environment where kids spend all day outside, there's only a small amount available, enough for the first three or four children in line. The rest get normal vision. In Singapore, there is a huge amount of nearsightedness, so the first few kids get a lot more, and there's still enough left for almost everyone to get some. In both cases, the kids line up in the same order, which is determined by their genetics. Regardless of the environment, the children at the front of the line (the most vulnerable) do worst, and those at the back of the line do best, but the amount of outdoor time determines how far down the line the problem is distributed. That's a common pattern in child development, called "gene-environment interaction," in which a genetic risk factor leads to a particular outcome in certain environments but not in others. The relationship between obesity genes and the modern environment works much the same way.

Such interactions provide a third potential explanation for the difficulty in finding obesity genes, the idea that some genes lead to obesity only under certain conditions. Some genes modify body weight across its entire range from thin to fat, while others influence only the chance of having a high body weight. In particular, many genes that affect obesity in mice lead to weight gain only on a tasty diet. For example, the odds of obesity in people are increased by some variants of G-protein-coupled receptor 120, which regulates appetite and fat storage in response to fatty acids.[33] In mice, mutation of this gene does not matter when the diet is low in fat, but animals on a high-fat diet gain more weight if they have the mutation. Finding genes like that would require separate genetic studies in populations that eat high-fat or high-sugar diets—or under whatever conditions activate the gene's effect on body weight.

Like nearsightedness, environmental influences on weight also mostly affect the genetically vulnerable, although we understand the details of that process only in rare cases. Fitness gains on a standardized exercise program vary from one person to another, largely because of differences in their genes.[34] When identical twins, men in their early twenties, were fed a thousand extra calories per day for about three months, each pair showed similar weight gains.[35] In contrast, the gains varied across twin pairs, ranging from nine to twenty-nine pounds, even though the calorie imbalance was the same for everyone. An individual's genes also influence weight loss. When another group of identical twins burned a thousand more calories per day through exercise while maintaining a stable food intake in an inpatient facility, their losses ranged from two to eighteen pounds and were even more similar within twin pairs than weight gain.

Take a moment to think about that. Some people's bodies resist weight loss so well that burning an extra thousand calories a day for three months, without eating more, leads them to lose only two pounds. The "weight loss is just math" crowd we met in the last chapter needs to look at what happens when their math is applied to living people. (We know what usually happens: they accuse the poor dieter of cheating, whether or not it's true.) If cutting 3,500 calories equals one pound of weight loss, then everyone on the twins' exercise program should have lost twenty-four pounds, but not a single participant lost that much. The average loss was only eleven pounds, and the individual variation was huge. Such differences can result from genetic influences on resting metabolism, which varies by 10 to 15 percent between people, or from differences in the gut. Because the thousand-calorie energy imbalance was the same in both the gain and loss experiments, this twin research also illustrates that it's easier to gain weight than to lose it.

Genes and Behavior

Of course most people do not diet or attempt to maintain their weight loss in an inpatient facility where someone else controls their access to food. That's important because genes influence susceptibility to the behaviors that contribute to weight gain. Patients with identified obesity-susceptibility gene alleles lose less weight after bariatric surgery than patients without those alleles.[36,37] As people grow from childhood to adolescence to adulthood, the heritability of obesity increases.[38,39] Presumably that happens because when people gain more control over their environment, they modify it to match their genetic tendencies. For example, a person might choose to live in the city, which requires a lot of walking, or in the suburbs, which requires a lot of driving, depending on genetic predispositions that affect the amount of physical activity that feels good.

Genes also strongly influence eating behavior. The correlation between spouses in calories consumed per day is 0.31 (where zero is no relationship and one is a perfect relationship), while the correlation between adult identical twins living apart is 0.69. Genetic heritage accounts for 60 percent of individual variability in emotional eating or in control of eating by willpower.[40] For instance, four-year-olds eat more of a snack offered after a meal if they have two copies of the higher-weight allele of the FTO gene mentioned earlier.[41] The higher-weight FTO allele also makes people more likely to eat calorie-dense foods with a lot of fat. In mice, a number of known genes increase vulnerability to overeating on a high-fat diet and a few increase overeating on a high-carbohydrate or high-sugar diet.[42] Because many more studies have tested the response to fat, though, this estimate is probably biased. Based on human behavior, I'd guess that researchers will eventually identify genes that increase vulnerability to overeating any type of highly palatable food.

Because genetic predispositions interact with environmental opportunities, the genes that matter for weight regulation in one environment may make no difference under other circumstances. Researchers had a chance to test this idea for the FTO gene by looking at participants in the famous Framingham Heart Study, which followed thousands of middle-aged people in a Massachusetts town starting in 1948 to look for risk factors that predict the later development of heart disease.[43] The study, which has produced about 1,200 research publications, continues with the children and grandchildren of the original participants. The FTO researchers compared people who were born before versus after 1942. Among people with two copies of the lower-weight allele, birth year made no difference to weight; those people were thin no matter when they were born. People with two copies of the higher-weight allele were heavier than their peers no matter when they were born, but the effect of the gene was larger in the group born after 1942. People with one copy of each allele were thin if they were born before 1942, but were susceptible to weight gain if they were born after 1942, into an environment that made it easy to gain weight. That dependence on circumstances helps to explain the increase in obesity in the modern environment.

Many thin people find that keeping their weight down is not too difficult, though it requires a bit of care. Extrapolating from their own experience, they tend to assume that deliberate weight control is possible, and therefore obesity must be caused by a lack of willpower. The lesson from genetics is that the strength of the impulses that willpower is called upon to control varies across individuals. This variation depends in part on genetic heritage, as expressed through the characteristics of the energy-balance and reward systems in particular brains. Thus, the differences in how easily individuals gain weight are much more likely to be due to biological characteristics than moral failures. If we want to understand someone else's weight gain, we need to imagine ourselves living not only in their circumstances but also in their brains. Genes can provide the motive to gain weight, but only if the environment provides the opportunity. To see how that process works, we must look beyond the individual.

10

Follow the Money

Edward Bernays, Freud's nephew, earned a reputation as the father of public relations. Bernays repeatedly changed public opinion by linking the satisfaction of unconscious emotional needs to the products he represented. In 1929, the American Tobacco Company hired Bernays to break the taboo against women smoking, then considered a habit of loose women.[1] After consulting with a psychoanalyst, Bernays decided that cigarettes symbolized male power, so he organized the "Torches of Freedom" campaign to encourage women to smoke Lucky Strike cigarettes. He got a list of debutantes from *Vogue* and convinced them that smoking cigarettes on Fifth Avenue during New York City's Easter Parade would strike a blow for women's rights. The press found the combination of celebrity and controversy irresistible and provided international coverage. Needless to say, smoking did not increase women's freedom, but linking the concepts allowed Bernays to open a new market for his clients.

Like the tobacco industry, the food industry has a problem. So does the diet industry. Both of them would rather have their customers believe that we're the ones with the problem—we're too weak willed to control our appetites and refuse to take responsibility for our weight. So far, the companies have done a good job at selling that message. It's a clever way to distract us from their problem, which is that they need to persuade individuals to buy their products, but they can't afford to admit where the profits lie in the industry. How many people would sign up for a diet program or buy a weight-loss book if they knew that the business model depends on repeat customers who come back after they've regained the weight they lost the previous time? Who would feel good about buying their family a nice snack of Hyperprocessed Heart-Attack Crisps if the package were labeled accurately? Yet diet plans don't make money by making people permanently thin, and food companies don't make money by selling crunchy fresh apples.

Both industries have huge marketing and lobbying budgets, so they have a strong voice in the conversation about their products. In 2014, the U.S. restaurant industry sold more than $683 billion worth of food, according to the National Restaurant Association. Grocery stores added another $620 billion, according to the Food Marketing Institute, with more than half of those sales from the ten largest food companies in the United States. In 2009, according to the Federal Trade Commission, the food industry spent $1.8 billion marketing to children alone, tying their products to popular movies, TV shows, cartoon characters, and video games to persuade children to pester their parents for unhealthy processed food.[2] Advertising for fast food, sodas, and breakfast cereals accounted for $1.3 billion of that budget. Children see forty thousand food ads per year, mostly for foods high in fat, salt, and sugar. The marketers have apparently read the research showing that early eating experiences have lifelong effects on food preferences.

The diet industry is smaller, with more than $60 billion in sales in 2014, according to Marketdata Enterprises, though we shouldn't call it a separate industry. In addition to diet clinics, some of which are owned by food companies, that sales number includes Diet Coke and Lean Cuisine, which are made by big food conglomerates. Unilever owns SlimFast, alongside brands like Ben & Jerry's. Nestlé diversified its processed-food empire in 2006 by purchasing Jenny Craig, which it sold in 2014. Heinz owned Weight Watchers from 1978 to 1999 and still manufactures the company's food products. The industry also includes more than 100,000 weight-loss books listed on Amazon in February 2015, whose publishers hope we won't notice that the average self-help customer buys a new book every eighteen months. The weight-loss industry wouldn't have so many repeat customers if its products worked.

Many of us spend decades responding to food-industry marketing that influences us to eat unhealthy food and gain weight, alternating with diet-industry marketing that promises to help us lose weight and get healthy, without realizing that we are listening to the same people. To restrict our food intake, we would need to pay attention to only one side of the mixed messages. For our own sanity, it seems important to realize that much of the weight-control and health advice out there is not offered in good faith.

Why Dunkin' Donuts Doesn't Sell Kale

The food industry's fundamental problem is that its companies need to win a fight with the brain's energy-balance system. One way to grow profits is to increase consumption, but that approach requires persuading people to eat more calories than their bodies need to maintain their current weight. Another way to boost the profit margin is to increase the amount of processing, so that foods can be made from mass-produced soybeans, corn, and wheat, yet still taste good enough to bring consumers back for more. Neither approach is good for the eater. Processing makes food less nutritious by stripping out fiber and nutrients, replacing only a fraction of the nutrient losses through vitamin enrichment.

Removal of fiber increases calorie density, making food easier to eat and interfering with some of the ways that the energy-balance system senses that we've had enough.[3] Liquid calories like the energy in soft drinks are particularly likely to be underestimated.[4] Similarly, studies in rodents suggest that the energy-balance system may compensate less effectively for the calories in sugared drinks compared with solid food.[5] For the companies, that's a feature, not a bug, as it helps remove one of the barriers to overeating (eating more than the body wants, an amount that is different for each person) and lets them sell more food.

A study in mice suggests that processed food additives may change the gut bacteria population in a way that promotes obesity and metabolic disease,[6] although we don't yet know if this work applies to people. In foods like ice cream and salad dressing, emulsifiers bind water and oils together, which lets these products sit on the shelf without separating. Emulsifiers can also disrupt the mucus barrier that lines the intestine, which keeps the bacteria inside the gut and out of contact with the immune system in healthy animals. The study found that two common emulsifiers, carboxymethylcellulose and polysorbate-80, in drinking water at low concentrations, changed the relative abundance of bacterial species in the gut and allowed them to penetrate the intestinal barrier. These changes led to inflammation, overeating, weight gain, and poor blood sugar regulation. Transplantation of gut bacteria from the affected animals into germ-free mice produced the same effects, indicating that they were indeed due to the gut bacteria changes. These emulsifiers also produced bowel inflammatory disease in mice that are genetically susceptible to the disease. If emulsifiers have similar effects in people, the rise of processed food might have contributed to the increases in all these conditions over the past half century.

Processed foods, which now make up three quarters of all food sales in the world, contain cheap ingredients like hydrogenated oils and sugar under a variety of names, from high-fructose corn syrup to reduced fruit juice. By adding more fat, salt, and sugar, food chemists persuade consumers to eat food that requires little chewing or digestion, provides a strong sugar spike in the blood, and is linked to diseases like diabetes and heart disease. Processed foods include not only chips and Cheetos, but cereal, crackers, bread, and pizza—indeed, almost everything bagged or boxed in the grocery store. Educated label readers can recognize processed foods through their long lists of ingredients. They also take a long time to spoil. Trans fats, which contribute to heart disease, are found in processed foods. Under pressure from activists, many food companies have replaced them with other processed fats whose health effects are unknown.

The switch to processed food was accompanied by population weight gain. From 1900 to the 1960s, physical activity declined as mechanization made work easier, but calorie intake decreased at the same rate, as the brain's energy-balance system continued to do a fine job for most people. This balance changed with a striking increase in fats and refined carbohydrates entering the food supply starting in the 1970s, as fast food became an increasing part of life in the United States. From 2 percent of children's calories in the late seventies, it had risen to 10 percent in the nineties. The major increase in obesity occurred from 1980 to 2000, which is also when the sales of processed foods went up dramatically.

During that period, food advertising increased by 60 percent, from $506 per person to $816 per person in inflation-adjusted dollars. Food marketing works.[7] That's no surprise, considering that companies are willing to spend $11 billion annually on it. Seeing ads makes us think more about particular brands, prefer them more, eat them more, and eat more in general. Although obesity is strongly associated with television viewing, watching noncommercial television or videos has no effect on children's body size five years later, suggesting that the association is caused by watching ads, rather than by sitting on the couch.[8] Between 1996 and 2005, 69 percent of movies contained food or beverage product placement—mostly snacks, sugary drinks, and fast food—an average of eight items per film.

Those decades have also been an amazing time for food science, with the introduction of new products peaking at more than sixteen thousand items in 1995 alone. In 1930, an entire grocery store contained only about a thousand items. By 1976, that number had reached eight thousand, and by 2000, it had soared to forty thousand. Portion sizes increased through the 1980s and 1990s as well. Even the current "medium" or "small" sizes are bigger than servings once were. In addition, food marketing has moved into schools since 1980. By 2000, about a quarter of middle and high schools sold brand-name fast food. It's unlikely to be a coincidence that between 1977 and 1996, the average food supply per person in the United States increased by 342 calories per day.

We're all confused about what we should eat these days, which makes the jobs of processed food marketers easier. Is it carbs that are poison, or is it fat? Should we be trying to eat smaller portions of our regular food or just skip the snacks? Are white foods bad? Would a juice cleanse help? Atkins? Paleo? Vegetarian? Intermittent fasting? Consumers now rate nutrition lower than taste or cost as a driver of their food decisions, and about equal to convenience.[9] In the confusion, it's easy to overlook how much of the food we eat is highly processed. Still, most nutrition researchers agree that processed foods are damaging to health.

Epidemiology strongly suggests that processed food, especially refined carbohydrates and added sugars, leads to diabetes. Added sugars made up 9.2 percent of calories in the U.S. diet in 2001, with the highest intake in people under forty. One study followed more than three thousand young adults for fifteen years and found that people who visited fast-food restaurants more than twice a week had twice as much increase in insulin resistance and gained ten pounds more during the study as those who went less than once a week.[10]

The Industry Strikes Back

Given the evidence that processed foods cause health problems, public health advocates have criticized food companies for selling them. To fight this criticism, these companies have taken several pages out of the tobacco industry's playbook, involving many of the same lobbyists and executives. First, they blame the consumer. The industry is only selling this stuff because people are buying it. If customers wanted healthy food, the companies would sell it. Anyway, the lobbyists say, there's nothing wrong with processed food as part of a balanced diet that also includes nutritious food and exercise. (Of course, making sure all our food is nutritious is likely to improve health even more.) They push the idea that obesity is caused by a lack of willpower, while overlooking the amount of time, money, and effort that the industry has put into challenging our willpower through food design and marketing.

These messages are spread by astroturf organizations like the Center for Organizational Research and Education (formerly the Center for Consumer Freedom), which is funded by food companies and run by a public relations firm that also works for tobacco companies. Food-company front groups even hire people to run social media campaigns and write negative comments on articles that disagree with their positions. They also promote the idea that healthy food doesn't taste good, which is much more prevalent in the United States than elsewhere in the world.[11] By popularizing these ideas, which are attractive in our rugged-individualist country, they hope to reduce public support for government actions to improve the food supply.

The individual freedom that the food industry promotes is an illusion. When I walk into a convenience store looking for a snack and find two hundred processed corn products, some peanuts, and perhaps an old banana, freedom of choice is not what I experience. People with little money may need to buy processed foods in order to get enough to eat. Processed foods are less expensive than whole foods because processed foods are made from government-subsidized crops, so the government already interferes with food production—in the companies' favor. The 2014 farm bill included crop insurance estimated at $90 billion, though it may cost more, and $44.4 billion in subsidies to protect against price decreases over the next decade.[12] Because the government pays part of the premiums for this insurance as well as paying for losses, agribusiness interests will benefit whether or not their crops fail. In contrast, the so-called specialty crops (otherwise known as fruits, vegetables, herbs, and tree nuts) will receive $800 million, less than 1 percent of the support for commodity crops, in block grants over the same period. The states can use this money for research, food safety, promoting consumption, expanding access, or a variety of other purposes, so it's not clear how much will reach farmers. In addition, in many areas, whole foods are not easy to get, especially for people without cars who live in neighborhoods that lack grocery stores.

Besides, eating whole foods requires cooking, which is more work for people who can barely find time to sleep, especially if they don't know how to cook. In 1965, almost all U.S. meals were cooked at home, mostly by women, with 92 percent reporting that they cooked. In contrast, by 2007, only two thirds of food was eaten at home, and just over half of people (68 percent of women, 42 percent of men) reported cooking.[13] Those who did cook spent an hour less per day on it, suggesting that they relied more on processed foods and preprepared ingredients. Half of the household food budget is now spent on eating away from home.

Second, the industry puts pressure on the government. The food companies spent more than $69 million on lobbying in 2014, according to the Center for Responsive Politics. What does it get for that kind of money? Leverage over the farm bill, the school lunch guidelines, and which foods can be purchased with federal food assistance, to name a few government decisions that went down that year. Perhaps more important in the long run, though, is the ability to influence federal nutrition recommendations. A major target is the Dietary Guidelines for Americans, which is revised every five years.

In the 2010 guidelines, the recommendations to eat more "fruits, vegetables, whole grains, fat-free and low-fat dairy products, and seafood" are straightforward. In contrast, the recommendations of what to eat less often are written in the language of nutritionists, which most people don't speak, telling us to cut back on "saturated fats, trans fats, cholesterol, added sugars, and refined grains." Why not phrase it as "red meat, processed foods, sweetened drinks, and white flour"? The food lobby understands that clear advice is more likely to be acted upon, so they push for recommendations in technical language. The preliminary report for the 2015 Dietary Guidelines was released in February of that year for public comment. It contained some major policy changes, such as lifting restrictions on cholesterol consumption and advising in plain language that people should eat less red and processed meats. Lobbyists for the meat industry geared up for action.[14]

In a less subtle move, in 2003, lobbyists for the Sugar Association (yes, there's a group that lobbies for sugar) got senators from sugar-producing states to pressure the federal government to pull funding from the World Health Organization if it published a report that suggested limiting sugar consumption to promote health. The Department of Health and Human Services sent a critique of the report that was largely written by industry lobbyists, and the final report made no mention of limiting sugar. Nutrition researcher Marion Nestle tells the detailed story of food-industry lobbying in her book *Food Politics.*

Third, industry lobbyists confuse the research literature and the journalists who report on it. (For the record, the only money I've received for writing about this topic is from my book's publisher.) Deliberate efforts to emphasize disagreements among scientists can reduce the public pressure for change by creating a false perception that the research linking hyperprocessed food to diabetes and heart disease is controversial and uncertain. Unregulated health claims like "smart choice" or "good for you" on new products that are slightly more nutritious than those they replace can improve a company's reputation without affecting the bottom line. That strategy is especially valuable because consumers eat more of products they consider healthy, whether or not that perception is correct. These "healthier" products are often much less nutritious than unprocessed foods.

The Center for Organizational Research and Education attacks researchers who argue for regulation, calling Marion Nestle "anti-pleasure" and "one of the country's most hysterical anti-food-industry fanatics" on its Web site. As its director, lawyer and lobbyist Richard Berman says, "Our offensive strategy is to shoot the messenger." This strategy includes referring to people who are trying to promote the availability of food options that our grandmothers would have recognized as "radicals." I guess that makes the companies who have completely reinvented our food supply in two generations out to be the conservatives. Anyway, it's easy enough to understand why people end up confused.

Much of the money for nutrition research comes from food companies, which raises concerns about conflicts of interest. A meta-analysis found that studies sponsored by food or beverage companies are four to eight times more likely to reach conclusions favored by the sponsors than those that are independently funded.[15] As I was writing this chapter, I received a press release for a new research center at the University of California, Davis, the Innovation Institute for Food and Health, in partnership with the Mars Corporation. Mars makes candy, alongside a few other foods, and it supports nutrition research. The Academy of Nutrition and Dietetics, which advertises itself as the world's largest organization for food and nutrition professionals, is sponsored by both Coke and Pepsi, along with food giant Unilever. Similarly, obesity researchers receive money from companies that make weight-loss drugs, which benefit financially from exaggerated claims of the harms of obesity and the odds of lasting weight loss.

To improve their public image, food companies also invest in initiatives to combat obesity. Their preferred approaches are to promote physical activity and to emphasize individual food choices. The Healthy Weight Commitment Foundation is a "CEO-led organization" with the mission of reducing childhood obesity by "helping people to achieve a healthy weight through energy balance." Its efforts to educate children from preschool through fifth grade about the importance of physical activity and a balanced diet are unlikely to be any more successful than the many previous efforts of this sort.[16] Children at that age are developing executive function, the ability to choose behaviors that move them toward a goal, but they do not yet have adult willpower or impulse control. Adults should not burden preschoolers, who are just learning to dress themselves without help, with the responsibility to consider whether their meals match their physical activity. Most adults are terrible at the deliberate control of eating, so it's ridiculous to expect children to do it. Such education squanders the major advantage that young children do have: at age three, most of them haven't yet learned to eat when they're not hungry.[17] Teaching children to ignore their hunger reduces the influence of the brain's energy-balance system, making them more vulnerable to food marketing.

This foundation also notes that its members have reduced their food production by 78 calories per person per day since 2010, largely because 82 percent of their sales growth was in lower-calorie products. The nutrition content of those products is not mentioned, though the Web site of one large member, Unilever, notes that 31 percent of its products meet or exceed (unspecified) nutrition guidelines as of February 2015. If the population eats fewer empty calories, that might reduce weight gain, but it is unlikely to result in weight loss for all the reasons discussed earlier, even if those calories are not replaced by foods from other sources. For health, the evidence is strong that maintaining current weight on a nutritious diet with regular exercise is better for most people than eating smaller portions of processed food, the solution that food companies would prefer.

Weight Loss, Inc.

Experts agree that preventing weight gain is much easier than achieving sustainable weight loss, but that's not what the diet industry advertises. Well, the information is there in the fine print, thanks to federal regulators, but not where it's easy to notice. As we know by now, the usual results of dieting are dismal. Counting calories every day for life makes a person about seven pounds lighter, on average.[18] A typical Weight Watchers client has lost six pounds after two years. A loss of that magnitude wouldn't change the pants size of many people.

In 1993, the Federal Trade Commission charged that five weight-loss programs made false and unsubstantiated claims about the effectiveness of their products. The companies included Weight Watchers, Jenny Craig, and Nutrisystem. Among the false claims were statements that clients were typically successful in reaching their weight-loss goals and that they maintained their weight loss permanently. To settle the charges, the companies were required to add disclaimers, "For many dieters, weight loss is temporary" and "Results not typical," to their ad copy. They also promised not to claim long-term success based on studies that lasted less than two years. In 2002, the FTC released a report suggesting that little had changed.[19] The authors estimated that 40 percent of weight-loss ads made a claim that is almost certainly false, and 55 percent made a claim that is very likely to be false. The struggle to regulate the industry continues, with seven new cases filed against weight-loss products in 2014 as part of Operation Failed Resolution.

The most prominent organization that claims to advocate for obese people is the Obesity Action Coalition, "dedicated to improving the lives of people affected by the disease of obesity." By now readers may be unsurprised that its major funding comes from the weight-loss industry. Its top three funders, the Chairman's Council Platinum members, in February 2015 were Covidien, which makes weight-loss surgery equipment; Eisai, which makes the weight-loss drug Belviq; and Novo Nordisk, which makes diabetes medications. Weight-loss surgery interests are well represented among the lower-tier sponsors as well. The Obesity Action Coalition promotes weight-loss surgery as a "safe and effective treatment" for obesity, despite its known risks, while they do not consider any approaches to obesity that do not involve weight loss.

In addition to these attempts to influence patients, weight-loss companies also pay doctors and lobbyists to increase the market for their services. The companies that make diet drugs have spent $10 million to $11 million per year since 2011 lobbying the federal government. A recent focus of their activity is the Treat and Reduce Obesity Act, which would direct Medicare to pay for diet drugs—mostly for people in an age range (over sixty-five) where obesity is not associated with excess mortality risk. These companies also provide money to scientific societies. Perhaps coincidentally, several of them issued guidelines recommending more treatment of obesity. The Obesity Society, the American Association of Endocrinologists, and the Endocrine Society have collectively received at least $4 million from drug companies since 2011.[20] Finally, the federal Open Payments database, established as part of the Affordable Care Act, shows that more than $5 million went to individual doctors and researchers specializing in obesity.

The research literature on this topic is full of hidden (and not-so-hidden) biases. Few medical facts are documented more thoroughly than the long-term failure of deliberate efforts to control body weight for the vast majority of people. As the National Academy of Science Food and Nutrition Board sums it up, "Many programs and services exist to help individuals achieve weight control. But the limited studies paint a grim picture: those who complete weight-loss programs lose approximately 10 percent of their body weight only to gain two-thirds of it back within 1 year and almost all of it back within 5 years."[21] Yet the federal government continues to fund—and researchers continue to perform—short-term studies of weight loss, ignoring the predictable regain that will happen after they pack up their data and go back to the laboratory.

The idea that weight loss is a reasonable goal for most people continues to be presented uncritically in many scientific papers, and few researchers attempt to separate the health effects of weight loss per se from the effects of improved diet and exercise. That's important because the evidence that does exist supports the idea that fitness and a healthy diet can prevent or reduce many of the health problems associated with obesity, whether or not the person loses weight.

The public-relations departments of international corporations would not argue for personal responsibility as the key to avoiding obesity if they thought that would cause people to cut back on their products. They know their customers are busy, distracted, and influenced by marketing, and that the kids are asking for sugary cereal because they saw it on their favorite Saturday morning cartoon. That's not an accident. It's the intended payoff of all the money that companies spend placing their products front and center in our minds. When we accept the cycle of repeated gain-diet-regain as normal, blaming our diet failures on lack of willpower instead of recognizing the power of environmental cues and the brain's efforts to maintain a stable weight, we have allowed people who do not have our best interests at heart to set the rules of the game.

Part Three

A Better Way

11

Eat with Attention and Joy

Gordon Rosenberg, an eighty-nine-year-old walnut farmer, had come to learn mindful eating at the Esalen Institute in Big Sur, California—famed for its meditating hippies and hot springs overlooking the Pacific. So far, he was finding it difficult to tune in to his body's signals. "I'm never hungry," he whispered to his wife, Vicky. Instead, as a "charter member of the Clean Plate Society," he ate whenever food was in front of him. He had trouble holding on to snacks long enough to listen to the instructions for mindful-eating exercises. Hounded by voices from his Depression-era childhood, he proudly ate the core and seeds of his apples, and he never threw leftovers away.

At its heart, mindful eating means making conscious choices about what to eat. The focus is on aligning our bodies' needs with our values, instead of giving up control to rules, habits, or external cues that don't support our goals. By giving ourselves permission to savor whatever food we want at that moment, without guilt, we can be fully satisfied after we eat. This approach allows us to get more pleasure from food, while eating exactly the amount we want, and so bring the brain's reward and energy-balance systems into alignment.

The only way to learn how to do that is to "play with your food," as workshop leader Jean Kristeller put it, by experimenting with different eating patterns and observing the results with curiosity and kindness. Gordon's intense need to eat everything on his plate interfered with his experiments. Other workshop participants suggested alternatives to wasting the unwanted food, from giving it to a local homeless person to sending an equivalent donation to a famine relief organization, but Gordon was unmoved. By the end of the weekend, with great effort, he had managed to leave a single bite of food on his plate.

Your Brain Knows How Much to Eat

Rigid food rules train us to ignore hunger, which reduces the influence of the energy-balance system. Mindfulness provides a set of techniques that help us to get back in touch with our bodies, as we let the voices in our heads flow by without undue influence on behavior. Strengthening the body signals from hunger and satiety gives the brain's energy-balance system more influence over eating. By relying on this system, which never takes any time off from work, we're protected against overeating in the inevitable moments when willpower fails. People with a strong disposition toward being mindful of their physical and emotional experiences are less likely to have disordered eating behaviors,[1] are less likely to act on eating-disordered thoughts,[2] are thinner, are more physically active, and have better cardiovascular health in middle age.[3]

Mindfulness combines deliberate attention to the present moment with acceptance of the world exactly as it is. Both aspects are necessary, and most people find both to be challenging, at least at first. Acceptance doesn't mean the belief that the situation is the best it could possibly be. Instead it's the deep realization that this is the only reality that exists right now, so it makes sense to live in it without trying to push it away. Though these skills are often practiced during formal meditation sessions, they can also be learned as part of daily life. I didn't manage to stick to a formal mindfulness practice on a regular schedule until several years after I started paying attention to hunger cues informally.

In a common type of formal meditation, people focus on their breathing while letting any distracting thoughts or feelings come and go without reaction. Remembering that thoughts are just thoughts, not facts or judgments from the high court, can help meditators to maintain emotional balance in the presence of negative self-talk or anxiety. Each of us gets to decide whether to take our thoughts seriously or act on them. I've never found the common instruction to "empty your mind" at all helpful, as it's not accurate and it tends to get me worrying that I'm doing meditation wrong. Meditation doesn't so much silence the "chattering mind" as cultivate an open, curious, and accepting attitude toward it. In people who meditate regularly, the ideas, labels, and judgments produced by the brain seem less powerful or compelling—more like a parade that we can watch from the sidewalk, without joining the procession.

Deliberate concentration is an important aspect of mindfulness practice, no matter what we choose to pay attention to. People who live mindfully might focus on the sensations in their hands as they wash dishes, the sounds of the birds in the neighborhood on their daily walk, or the feelings of hunger and fullness in the body as they eat. Whenever people are actively paying attention to what's happening without passing judgment, they're practicing mindfulness. Although mindfulness originated in the Buddhist tradition, its practice is compatible with any spiritual beliefs or none at all. As with any other skill, the more people practice mindfulness, the better they're likely to get at it. Sitting in formal meditation provides a structured opportunity to practice mindfulness every day, which sharpens the powers of concentration and improves sensitivity to signals from the body.

Mindfulness practice trains us to deliberately direct attention and avoid being captured by intrusive thoughts, which has many positive effects on emotion regulation.[4] In the clinic, mindfulness-based therapies successfully treat anxiety and depression[5,6] and prevent relapses in people with repeated episodes of depression.[7] This effectiveness might be explained by changes in the brain.[8] In mindful people, the prefrontal cortex, an area important for self-regulation, is good at inhibiting the amygdala, a brain region associated with rapid emotional responses. Training in mindfulness improves people's ability to handle stress, even in tough circumstances like persistent pain.[9] For those patients, the acceptance and nonjudgment aspects of mindfulness may be particularly important to the effect.

Mindfulness practices can also increase awareness of our eating habits. Looking without judgment at how we behave around food makes it easier to see our actions clearly—and contrary to what we might expect, it also makes them easier to change, if that's what we decide to do. Inserting a pause for reflection into a cascade of habitual actions can open up space for mindful decisions about what to do next, eventually leading to the formation of habits that serve us better. Mindful eating improves the ability to experiment with different ways of eating, examining how each of them makes us feel, physically and psychologically, with curiosity rather than harshness. When meditators notice that they are making judgments (and everyone does), they don't try to resist the impulse or get angry with themselves. They just calmly and patiently direct attention back to the focus of concentration.

Attempting to resist intrusive thoughts, such as food cravings, drains willpower and often ends up making the thoughts stronger anyway.[10] Instructions to avoid thinking about pink elephants reliably produce a flood of thoughts about pink elephants, presumably because the cognitive control system's attempts to check whether the task instructions are being followed ends up generating such thoughts. Letting them flow past without judgment takes a lot less effort, once we learn how to do it.[11] Training in mindful acceptance reduced food cravings and the frequency of thoughts about food in one small study.[12]

Another aspect of acceptance is letting go of our ideas about the desired outcome of mindfulness practice. Striving toward a specific aim harms the larger goal of learning to sense the body's needs in the moment because it keeps people from watching their experience as neutral, nonjudging observers. When people are trying to steer mindfulness practice toward a foregone conclusion like "If I eat mindfully, I will eat less," then they're not experiencing the moment exactly the way it is. Mindful eaters do not always eat small amounts of food. Sometimes the body's hunger signals tell us that we want a big meal. The purpose of mindfulness is to allow people to eat as much food as their bodies want, no more and no less. Trust that the body knows what it needs. Thousands of generations managed to figure out what to eat without calorie-counting apps and diet books, and so can we. Trying to second-guess the body's needs through deliberate effort or to talk ourselves into eating less than we want puts us back in diet mode. By now we know how that story ends.

Listening to the Body

The idea that it's psychologically and physically healthier to eat in response to internal cues than to count calories goes back to the antidieting movement of the 1970s and 1980s. As eating disorders and routine temptations to overeat became increasingly common, many women realized that restrictive diets caused more problems than they solved and started looking for a better way. Dietitians Evelyn Tribole and Elyse Resch describe this approach as "intuitive eating" in their book of the same name. They define intuitive eaters as people who give themselves permission to eat whatever they want when they're hungry, eat in response to hunger and satiety, and don't eat for emotional reasons. Mindful eating differs from intuitive eating mostly in the set of techniques that are used to develop this ability.

From middle school[13] to middle age,[14] intuitive eaters weigh less than their peers and are protected against weight gain.[15] Intuitive eaters enjoy preparing and eating food more, and although they are less health conscious, they eat a more varied diet. They are less likely to binge[16] and more likely to exercise because it feels good.[17] One small study found that intuitive eaters have lower triglycerides and higher HDL (good) cholesterol compared with eaters who ignore internal cues.[18] Intuitive eaters also have fewer symptoms of eating disorders, are less depressed, have higher self-esteem, feel less pressure to be thin, and do not value thinness as an ideal body shape.[19] In contrast to the temporary effectiveness of dieting, which leads to worse results as time goes on, intuitive eating studies produce better outcomes with longer follow-up.[20]

As we might expect, intuitive eaters are better at sensing the internal state of their bodies, an ability that can be measured by a heart-rate monitoring task.[21] Participants are asked to silently count their heartbeats over an interval of less than a minute, without taking their pulse. Performance on this task correlates with the ability to sense signals from other parts of the body, including feelings of hunger and satiety, and with the ability to identify what emotions another person is feeling.[22] Better performance on this task also correlates with a lower weight. People who are good at identifying their own sensations tend to be good at regulating their behavior and emotions. In contrast, a tendency to avoid awareness of body sensations is associated with disordered eating and body dissatisfaction. For people who don't like their body, it's hard to concentrate on it or take care of it.

Practice can train the ability to perceive hunger and fullness sensations. One way to improve is through meditation, especially the body scan, in which a person deliberately pays attention to the sensations in each part of the body, one at a time. This practice can also help to decrease body dissatisfaction. Many guided meditation recordings include a body scan for anyone who's interested in trying it.

Training in mindful eating also helps people with binge eating and other eating disorders learn to eat normally.[23] Part of the definition of binge eating is that it continues well past the point when the body stops feeling good, so bringing awareness to this transition can help people avoid crossing that line. Binge eaters respond more than other people to external cues, like food availability, and are less sensitive to internal cues of hunger and fullness, which makes them vulnerable to binge triggers. Traditional diet programs typically make this problem worse by encouraging dieters to ignore their hunger and follow a cognitively controlled food plan, often leading to cycles of binge eating, followed by shame and renewed efforts at restraint. A more effective approach is to improve the ability to sense internal cues and self-acceptance by cultivating mindful awareness so we can eat with moderation and joy.

Categorizing foods as "good" or "bad" is rarely helpful because rigid food rules tend to lead to binges when they're violated. In any case, few foods are as demonic or angelic as we make them out to be. A person who was stranded on a desert island would live a lot longer on hot dogs than on spinach. Removing all the fat from the diet would damage health, as cell membranes could not be produced or repaired, and electrical impulses along the nerves would not conduct properly. Cholesterol is needed to produce sex hormones and vitamin D. Carbohydrates provide the only source of fuel that the brain can use, in addition to supplying energy to other tissues, which is important to support exercise. If possible, try to see foods as individuals with good and bad points, not as one-dimensional saints or sinners. That way of thinking is likely to lead to better dietary choices than an all-or-nothing approach.

Mindfulness training can help people move from eating that is controlled by external cues to intuitive eating, including those of us who just want to cultivate a more balanced and comfortable relationship with food. The program that Gordon Rosenberg attended at Esalen, called Mindfulness-Based Eating Awareness Training (MB-EAT), improves awareness of hunger and decreases both frequency and duration of binges in clinical trials.[24] The program includes a series of mindfulness exercises, in which participants are encouraged to pay attention to their experiences as they eat a variety of foods. That experience trains them to become aware of hunger, fullness, and taste sensations so they can maximize the pleasure they get from eating, to cultivate body acceptance and awareness, and to choose foods deliberately and pay attention to the results. Formal sitting meditations are used to teach mindfulness skills, including loving-kindness meditation to increase self-acceptance.

Perhaps most important, participants are encouraged to use mini-meditations throughout the day to calm and center themselves before making food decisions. In this technique, which takes only thirty seconds or a minute, people take a few deep breaths to relax and then mindfully consider hunger, other motivations for eating, and what kind of food (crunchy? sweet? savory? warm?) might best satisfy those needs before choosing whether and what to eat. The participants who regularly used the mini-meditation technique showed the most improvement in their eating patterns. Jean Kristeller, the researcher who designed the MB-EAT program, notes that clinicians who have been trained to teach the program often tell her that their own eating improved as a consequence of working through the material.

The goal of this program is not to stop using food for pleasure or to meet emotional needs but to learn to do it intentionally and effectively. People who find that they often eat when they're not hungry, to soothe emotions or avoid boredom or whatever, can work on finding more effective ways to cope with those needs. In addition, just accepting that we've decided to eat for comfort can lead us to be satisfied with less food. Trying to eat for comfort while experiencing guilt and shame produces little benefit, which leads some people to chase after the experience they wanted by eating more while remaining unsatisfied. Making peace with food includes recognizing that pleasure is a valid reason to eat. If we restrict ourselves to foods that we don't enjoy, we're likely to find ourselves poised to overeat at the next opportunity. Learning how to eat what we want on the first try is key to improving satisfaction and reducing struggles with food.

Other programs do not use formal mindfulness practices but still teach participants to eat when they're hungry and stop when they're full. One example is called Appetite Awareness Training. This program, which was initially designed to treat eating disorders and then extended to help anyone who agonizes over eating decisions, trains people in the skills needed to rely on internal eating cues. There is also a workbook for self-guided use.[25]

Simply reading about mindful eating is unlikely to change behavior. Only learning through experience can do that. Although anyone can read a book or take a training class to help, one of the main lessons of mindful or intuitive eating is that no one can tell other people exactly how to feed their bodies. The way to find out what a particular body needs is to get in touch with its signals and then do some experiments. If we approach this phase with curiosity rather than judgment, we can learn as much from the mistakes as from the successes.

How to Eat Mindfully

A few tactics can help you learn to become a more mindful eater. Begin meals at a moderate level of hunger because people who are starving tend to eat too fast, which makes it hard to tell when they're full before they're overstuffed. Along the same lines, try to eat more slowly than usual, which also provides more taste sensations. If possible, create a space to eat without distractions, at least for some meals and snacks. If an entire mindful meal is too much to manage, consider starting to eat with a brief silent ritual of tasting the first bite of each food carefully, experiencing its appearance, smells, flavors, and textures with gratitude. Pay attention to the way the body feels before eating, stop in the middle of each meal to check on hunger, and note how the body feels after eating. If the answer is that it feels stuffed and uncomfortable, congratulations—that's an important lesson about how to feel better next time. The more different kinds of food and eating patterns that we try, the more information we have to decide what we want to eat.

During the learning process, some people find it helpful to take notes on their feelings, while others rely more on their accumulated experience. The note-taking types might try rating hunger on a scale of one to ten before and after they eat. They might also want to record what they ate, the emotions they were feeling, how satisfied they felt after eating, and any environmental situations that seem relevant, such as being in a restaurant or eating while distracted. Looking back over these notes can sometimes help people to see patterns that weren't clear at the time. There's no need to keep notes on food quantities, especially if that tactic draws the focus back to weight loss, as it does for many recovering dieters.

In addition to hunger, try focusing on taste and how it changes. The first bite of food tastes better than the fifth bite for most people because of a phenomenon called taste-specific satiety. Responding to taste satiety naturally leads people toward a diverse diet, as they tire of eating the same things day after day. The lure of new tastes may explain why we find room for dessert even when we feel physically full. If we focus on being fully aware of the taste as it develops, we may find ourselves satisfied with a few bites instead of either fighting the craving or cleaning the plate out of habit. Indeed, we may find that the anticipation of certain foods is more pleasurable than the experience of eating them.

After I stopped dieting, I came to realize that some foods I'd craved as treats, such as doughnuts, didn't taste all that good to me when I tuned in to my sensations during and after eating them. Somehow I'd been taken in by the social signals telling me that everyone loves doughnuts, so I managed to convince myself that I had superior self-control because I didn't give in to those cravings often. Now I find it more strange that I sometimes did, because doughnuts never made my body feel good, and they taste like sugar straight from the bag—boring and too sweet. When I avoid filling myself up with foods that aren't enjoyable or nutritious, I can make space for my true pleasures in a balanced diet. High-quality ice cream, for instance, remains every bit as delicious in reality as it ever was in my imagination.

The learning process took most of a year for me. Persistent attention to hunger may not sound much more inviting than constant worry about counting fat or carbohydrate grams, but there's a big difference between the two approaches. The attention to hunger is temporary. People who've learned how to eat mindfully don't spend a lot of time deciding what to eat and how much. Once the ability to sense hunger has been practiced enough, it becomes a habit. Then we can use the constant vigilance of the brain's energy-balance system to our advantage, rather than fighting against it. Experienced mindful eaters are relaxed around food and rarely think about it when they're not hungry. They know what their bodies want, and they're not afraid to give it to them.

In contrast, many people who've spent a long time dieting have become convinced that they should not trust their judgment around food. Dieting teaches rigid rule following and obedience to external authority, habits that we need to unlearn if we are to take control of our own food choices. In the beginning, a common fear is that if we ate anything we wanted, we would gain a hundred pounds on cake and cookies, eating ourselves into an early grave. Some people do go through a phase of rebelling against their restrictions by eating formerly forbidden foods, but repeated exposure to those foods leads to habituation, reducing their appeal to normal levels. If that stage happens at all, it is transient and can be minimized by experimenting with trigger foods in smaller packages or outside the house. Usually it ends when people realize through their own experience that sugary or fatty foods don't continue to taste or feel good after they've eaten too much, and that their bodies are craving something else.

After a lot of dieting experience, it's easy to slip into that mentality when thinking about food, whether it's counting grams of fiber or avoiding certain foods. Until we give ourselves full permission to eat when we're hungry and stop when we're full, the cognitive control system is still in charge. That means the pressures that lead to binge eating may continue to bear down on us. Such unconditional permission can be scary, but the benefits of eating this way depend on giving up deliberate control of weight. Dennis Asbury said of his first experiment with eating mindfully that calorie counting "was like a life raft or a piece of driftwood I was holding on to in a big sea of rough waters. I was afraid to let go of that because I just knew that I was going to sink."

Being mindfully present with that anxiety, observing it without getting too caught up in it, can help us to change those patterns and relax around food. Effective treatments for anxiety center around exposure in a safe context, to help people learn that they can handle the situation that scares them. If we deny ourselves that learning opportunity by avoiding the foods that make us feel out of control, the anxiety they provoke is unlikely to go away. Whatever happens while experimenting, be kind about it. Beating ourselves up when things go wrong prevents us from seeing our behavior clearly, and that makes the lessons harder to learn. Making and correcting errors is the only way to learn anything new.

Mindful eating honors the needs of the body. To keep the focus on sensory experience during the transition period, avoid scales, calorie counts, or food plans of any description. The early stages of the transition to mindful eating also may not be a good time to focus on nutrition. Once the mindful approach has become comfortable and we can remain aware of the body's reactions to food, there will be plenty of time to consider ways to make healthier choices. Especially for experienced dieters, it's far too easy to fall down the rabbit hole of chasing a nutritional goal by limiting food choices. In the experimental phase, that approach is likely to be counterproductive. Recall that the brain's willpower system fails when it's overworked, while the energy-balance system remains reliable all the time. The purpose of this process is to put the energy-balance system firmly in control. That's not possible if we're second-guessing its decisions by counting, weighing, and measuring, as those approaches fall into the domain of the executive-function system.

Mindful eating may help people lose weight if they often eat beyond the point when they're no longer hungry, or it may not. Either way, this approach to eating is a powerful tool to maintain weight stability. Imagine needing to own only one set of clothes, which always fit, without deprivation or cognitive effort. If we teach ourselves to eat in a way that's relaxed and satisfying, our weight will stabilize within the range that the brain's energy-balance system defends and stay there for as long as we continue to work with the brain instead of fighting against it.

The challenge of finding ways to pay attention to food during our busy lives was a theme of the Esalen workshop. On the last morning, one of the participants was starting to slip out of a meditative frame of mind while packing his clothes. As he rehearsed all the tasks he needed to do as soon as he got home, he was startled out of his reverie by the realization that a raisin from the first mindful eating exercise of the course was stuck to the bottom of the shoe he'd just picked up. In that moment, he discovered his most important take-home message. He thought it would make a good T-shirt motto: Don't forget to stop and smell the raisins.

12

Sleepwalking Through Dinner

When the customers arrived to see the new Mel Gibson movie on a Saturday afternoon, they received a surprise. To celebrate the theater's one-year anniversary, or so they were told, they got a free soft drink and popcorn. The habit of eating during movies was so strong that most people overlooked a serious problem: the popcorn was five days old, stale enough that two customers forgot it had been free and asked for their money back.[1] People ate slowly, taking a handful or two and then pausing. Still they kept reaching into the bucket, over and over again.

The popcorn came with an ulterior motive. Brian Wansink's research group provided it as part of a study on the effects of serving size on eating.[2] As the customers left, they were asked to fill out a questionnaire, and the remaining popcorn in their medium or large buckets was weighed. Most people who received large containers denied that they had eaten more popcorn as a result, but they put away 173 more calories on average than people with medium containers, though both sizes held more food than they could finish. No one was eating for the taste because the popcorn was terrible. They also probably weren't hungry, as the movie started right after lunch. Instead they ate out of habit, mindlessly.

Avoiding situations that promote mindless eating makes it easier to eat mindfully. Regularly eating past the point of fullness can lead to irreversible weight gain, as the defended range inches up slowly to match the new higher weight. The modern environment makes that process as frictionless as possible. Between our own evolved tendencies to minimize activity and seek out tasty foods, and the marketing that is designed to take advantage of them, mindless eating has never been easier.

Food Cues Everywhere

How much of our eating is unconscious? A lot more than we realize, because the brain does most of its work behind the scenes. The important stuff in particular doesn't tend to be left under the control of our distractible conscious minds. We don't risk forgetting to breathe when we get caught up in a story or forgetting how to walk when we argue with a friend. Eating decisions are also automatic in many circumstances. External cues, from the design of plates to social norms, strongly influence most people's food decisions—especially controlled eaters who have learned to disregard their internal cues—and most of us are not aware of these influences.[3] Modifying our environment so these external cues push us to eat less is one way to decrease unconscious eating. Another way is to practice mindful eating. For most people, a combination of these two strategies is likely to be more effective than either alone, especially in the early stages of mindfulness practice.

Occasional overeating isn't a problem. Because food resources have been variable over our evolutionary history, the brain's energy-balance system is well prepared to deal with inconsistent supplies. In small children and other intuitive eaters, big meals are typically followed by small ones, so that average energy intake over a week or two remains constant. People who minimize the everyday temptations to eat without noticing can choose to overeat at a holiday meal or a fancy restaurant dinner without any consequences except feeling less hungry for the next couple of days. Weight gain happens when mindless eating consistently numbs us to the signals of fullness that are supposed to limit meal size.

Although a fantastic meal can enhance our life even if we feel overly full afterward, most of us would benefit from cutting out the stale popcorn in front of the television. Mindless eating isn't a pleasure, and it rarely involves nutritious foods either. Almost everyone claims not to be influenced by the factors that promote mindless eating, but all of us are.

There are starving children in the world, but whether we clean our plates doesn't matter to those children. In the United States, we waste about 40 percent of the food produced, more than twenty pounds per person every month, which has increased 50 percent since 1974. Much of that waste happens on farms, in food production plants, and in restaurants, but estimates suggest that 15 to 25 percent of food that comes home from the grocery store ends up in the trash.[4] Being a member of the clean-plate club, as many of our parents taught, doesn't solve that problem. Food is just as wasted if it's eaten by someone who's not hungry as if it's thrown away. Unsurprisingly, people who stop eating when they're full tend to weigh less than people who stop eating because the plate is empty. A better tactic is saving leftovers for later, when most of us will get more pleasure from them. If we bought less food or donated extra supplies to a food bank, that might make a genuine difference to people who don't get enough to eat.

People are more likely to stop eating when they're full if they arrange the food environment to make stopping easier. We all rely to some extent on habits or external signals to decide when to finish eating, so they might as well be organized to help. An empty dish is a strong cue that the meal is over, as researchers discovered by secretly refilling soup bowls from underneath, leading eaters to consume 76 percent more food while estimating that they had barely increased their calorie intake.[5] Perceived variety increases consumption as well, with people eating 43 percent more M&M candies from a bowl with ten colors, which all taste the same, instead of seven colors.[6] Mindless eating can be triggered by seeing or smelling food. People serve themselves smaller portions in taller, narrower glasses or smaller plates or bowls, and when using smaller serving spoons.[7]

Over the past century, not only have portion sizes increased but plates have gotten larger as well. Our great-grandparents ate from dinner plates the size of today's salad plates. Eaters can prompt themselves to notice when they're full before they eat too much by using smaller plates and smaller servings, and by keeping the second helpings in the kitchen instead of on the table. If the plate has been empty for a few minutes and you're still hungry, then by all means go get another serving. The point of this approach isn't to trick yourself into eating less than you really want, just to protect against accidentally stuffing yourself.

Another external cue that many people use to decide how much to eat is food volume. In short-term studies, researchers find that we tend to eat a certain quantity of food for breakfast, lunch, or dinner, and people who get less than their usual amount feel unsatisfied. That's probably a combination of expecting a certain number of bites and expecting the stomach to feel the typical way after a meal. This mental shortcut creates a tendency to overeat energy-dense foods like candy or steak, which can contribute to weight gain if it happens often. One tactic to avoid eating more than the body wants of energy-dense foods is to eat slowly, giving the brain's fullness sensors a chance to keep up. Another approach is to start the meal with a soup or salad and save the energy-dense food for last. People who cook their own meals can reduce the energy density of some foods by blending in vegetables or serving them alongside.

Modifying the effort required to obtain food, just by a tiny bit, has one of the strongest unconscious effects on how much of it is eaten.[8] The ease with which food is available has increased enormously over the past century and is a major contributor to unconscious eating. Workers eat fewer candies when they have to walk six feet to the candy dish than if it's placed on their own desk. Cafeteria patrons eat half as much ice cream if they have to open the freezer door than if it is left open. People eat fewer unshelled than shelled almonds, and they eat less if they have to open multiple small packages than one big package. The lessons for the home are obvious: keep sweet and fatty foods at the back of the cupboard or refrigerator, or move them to the basement or laundry room, while making healthy snacks like nuts or fruit easy to find. Don't keep any food (except perhaps fruit) on the counters where it's visible anytime someone walks by. Cut up vegetables before putting them away so they're convenient to eat as snacks. Leftovers can be a problem in this context, because they're easy to pull out and eat. If they're too tempting, freezing them can provide enough of an energy barrier to make people stop and think about whether they're really hungry before eating.

A great way to reduce mindless eating is to avoid mindless shopping. Once food comes into the house, people are more likely to eat it. That's not a recommendation for the empty-kitchen diet, which would likely lead to eating a lot of take-out and restaurant food. Instead, people can fill their kitchens with the foods they'd like to be eating more often, bought in larger quantities. Buy small bags of potato chips and large bags of green beans. Set a target to buy a particular amount of vegetables or fruits, or no more than one box of crackers. To start with, choose foods that the family already eats, not a bunch of aspirational ingredients that no one knows how to prepare. When the fridge is full of weird food that no one has any idea what to do with, many people end up going out for pizza. Include some healthy snacks that are easy to grab and don't require much preparation. One place where people are especially likely to eat whatever's convenient is at work. People who pack their own lunches are healthier than people who eat out, and the ones who eat from vending machines are in the worst shape. Bringing good food to work makes it easier to do the right thing.

People also eat more when they're comfortable. Restaurants invest in soft lighting and mellow music because a pleasant dining experience tempts us to linger over dessert or another drink. Asking the waiter to wrap up the leftovers as soon as we feel full, or ordering a cup of coffee after a meal, can help to avoid mindless eating while allowing us to socialize. Eating with friends increases the length of meals and the amount of food consumed, while we tend to eat less when we're nervous, as on first dates or job interviews. Other people can also define social norms, suggesting how much food it would be appropriate for us to eat under the circumstances, which can increase or decrease eating, depending on our choice of friends or colleagues. Most of us prefer not to be the only person at the table who orders dessert, but we might eat some, even if we're not hungry, if everyone else does.

Distractions, like watching television while eating, make it harder to notice internal cues and thus let external cues control consumption. People who eat in front of the television report being less hungry while eating than those who eat without distractions. Eating under these conditions is typically habitual, so make sure to choose small portions or healthier snacks and keep the refills hard to obtain. Never eat snacks directly from a large package while distracted. People who watch a lot of television are heavier on average than those who don't watch it, probably through a combination of mindless eating and exposure to food marketing. For the biggest decrease in mindless eating, try not to multitask while eating at all.

Avoiding mindless eating is an excellent strategy to prevent the creeping weight gain of a pound or two a year that most adults experience in the modern environment, often without knowing what caused it. That's a big deal, as preventing the defended weight range from inching up would leave the average person fifty pounds lighter by age seventy. Yet despite the claims of experts like Brian Wansink (whose book *Mindless Eating* is otherwise excellent), avoiding mindless eating will probably not make most people lose a large amount of weight because it won't lower the defended weight range.

Like other lifestyle changes, eliminating mindless eating can get people's weight down to the low end of their defended range, without feelings of deprivation. In some cases, people who have gained weight by eating beyond hunger repeatedly may discover that their defended range has not increased to match, allowing them to lose the weight when they stop overeating. When anyone's weight drops below the lower end of that range, though, the brain's energy-regulation system will promote weight regain, using tactics that are just as unconscious as the ones that food marketers use to induce us to gain weight.

Raising a Mindful Eater

Mindless eating typically starts in childhood. Many parents end up teaching their children to rely on external cues instead of hunger, which increases their later vulnerability to food marketing and emotional eating. Young children's eating is driven by their internal cues of hunger and fullness. Infants change the amount of milk they drink based on its calorie content. At age three and a half, children eat exactly the same amount of food regardless of how much they're served, but by age five, consumption increases steadily with portion size, as it does in adults.[9] Serving a smaller portion helps five-year-olds to eat the amount that they want, as does putting the food on a child-sized plate. The intuitive eating skills of young children allow them to regulate their food intake accurately across weeks, despite a lot of variability in the size of individual meals.[10] Watching a toddler eat almost nothing at some meals bothers a lot of parents, though. Instead children are taught to respond to external cues like Grandma's urging to have more of her special meat loaf or the requirement that they clean their plates, even if they aren't hungry. Parental anxiety about overeating, expressed through rigid limits on serving size or reluctance to provide second helpings on request, can also teach children to disregard their hunger signals. Unless a child has a genetic defect in the brain's energy-regulation system (characterized by constant, voracious hunger), parents can trust internal cues to regulate that child's calorie intake without any second-guessing.

Parents may alternate between pressuring their children to overeat and worrying that they might be eating too much. The interactions between mothers and daughters around food are often fraught. Mothers may have their own anxieties related to eating and body image, and they suspect that they're likely to be blamed if their children grow up to be obese. When these anxieties are expressed by attempting to limit how much food the child is allowed to eat, a damaging cycle can be set into motion. Children whose food is restricted are more likely to eat for emotional reasons, a behavior that usually appears after age seven.[11] Food restriction also increases children's preference for the forbidden food and leads them to overeat it whenever it's available, especially among boys.[12] This pattern is reminiscent of the effects of dietary restriction in adolescents and adults. Because parents can't control their children's environment indefinitely, the restriction strategy is doomed to fail.

One long-term study followed 140 families with young girls from the ages of five to nine.[13] On average, the girls' tendency to eat when they weren't hungry increased with age, from starting levels that were similar across all groups. Food restriction at age five predicted eating without hunger at ages seven and nine, whether the girls had been normal weight or overweight at five. The largest increases in eating without hunger were seen in girls who were already overweight at age five and experienced food restriction, perhaps because weight concerns motivated more intense efforts at restriction for those girls. But restriction is no more successful at controlling weight in children than it is in teenagers or adults. Instead it causes the same counterproductive behaviors—eating in response to external cues and overeating when food is tasty—that lead to mindless eating and weight gain. Girls whose food is restricted are more likely to become overweight over the next few years than girls who are allowed to eat as much as they want.[14]

The most successful interventions to reduce obesity or promote healthy behaviors in children do not focus on weight or necessarily involve the children at all. Instead they help parents and schools to provide an environment with healthy food and opportunities for physical activity. Regular weigh-ins or expressions of concern about a particular child's weight are more likely to lead to an eating disorder than to prevent future obesity. Indeed, they may increase the odds of obesity.

To test people's ability to sense hunger and fullness, researchers ask study participants to eat a serving of food or drink and then allow them to eat as much as they want of a snack or meal. As the calorie content of the initial serving is varied, people who are good at relying on internal cues adjust their later eating accordingly. People who respond to external cues tend to eat about the same amount in both conditions. The adjustment, called caloric compensation, becomes weaker with age and is less effective in heavier children. One study of siblings showed that five-year-olds compensated almost perfectly for the energy contents of the initial serving, while normal-weight older children tended to overcompensate by eating fewer total calories when the initial serving was large, and overweight older children undercompensated by eating more total calories when a bigger initial serving was provided.[15] Another study, of four- to seven-year-old twins, showed that mothers were more restrictive with the heavier twin of a pair, who was less good at caloric compensation.[16] These studies suggest that food restriction leads to reliance on external cues and weight gain, rather than restriction merely being a parental response to genetically driven overeating in their children.

Food restriction in children is also tricky from a nutritional perspective. Low-calorie diets may prevent them from growing properly, and low-fat diets are associated with a lack of important nutrients. Because they're growing, children need more calories than adults per pound of body weight. Young children also need to eat a lot more fat than adults do. The National Institutes of Health recommends that children under two should get 50 percent of their calories from fat, and children older than two should get 25 to 35 percent of calories from fat. Complicating things further, it's hard to be certain how much a child should weigh because kids often put on extra fat before a growth spurt. Relying on the child's hunger and fullness to regulate food amounts is a good way to manage this complexity.

That doesn't mean that children should make all their own food choices, lest they subsist for the rest of their lives on candy. Nearly all children like salt, sugar, and fat, which are the innate food preferences that do not rely on experience. Most other food preferences form early in life, beginning in the womb and continuing through roughly age two. To protect children against overvaluing sweets, give them early experiences with a variety of foods, as flavor preferences learned in infancy can last for years. Children prefer foods that are familiar to them, including tastes from the mother's diet that were present in amniotic fluid and breast milk, so even mothers who don't like vegetables might want to make the effort to eat them during pregnancy. If a breast-feeding mother consumes garlic oil or carrot juice, babies are less likely to react if the same flavor is added to a bottle. Babies who are fed soy formula (because of an allergy to cow's milk), which is bitter tasting, are more tolerant of bitter flavors and more likely to enjoy bitter vegetables at age five than their peers who drank milk-based formula.[17] In general, a good division of responsibilities is that the parents decide what foods are available and when, while the child decides how much to eat.[18]

Although children do need to eat their vegetables, making that a focus of family conflict is unlikely to produce the desired results. In contrast to our culture's expectation, allowing three- to six-year-old children to decide how much they want to eat is associated with increased consumption of fruits and vegetables.[19] For the vegetables that children already like, serving larger portions is likely to increase how much they eat. Trying to expand a child's range of acceptable foods to include more healthy choices is also worthwhile, though it does require some effort.[20] Allowing children to help cook something often makes them more interested in eating it. Slowly increasing familiarity by asking them to take one bite of a questionable food on a dozen occasions is more effective than requiring them to eat a full serving on the first try. Another way to improve children's liking for a vegetable is mixing it with a food they enjoy, such as milk or even sugar, which eventually leads them to like the vegetable alone. If necessary, young children can be given vegetables by pureeing them and mixing them into other foods.[21] All these approaches work for adults who dislike vegetables too.

Another project that's worthwhile with young children is to increase the variety of foods that they enjoy. At age one, most kids aren't too fussy about what they shove into their mouths (to put it mildly), so offering a wide selection of foods at that age is likely to expand the range of tastes that they like in adulthood. Around age two, most children become much more selective, so it's best to start early. Seeing someone else make an unpleasant face while eating a food is often enough to persuade kids not to try it. Social learning has been observed in young babies, and it has a big role in food preferences, with children being predisposed to like foods that are offered by friendly adults.

That learning is a double-edged sword for people with food anxieties, which can lead to behavior during meals that passes the anxiety along to the next generation. To avoid burdening children with unhealthy food habits, don't use food as a reward or punishment ("You have to sit there until you finish your broccoli"). Also try to be careful about labeling foods as "good" or "bad." The problem with labels is that people tend to see things in black and white, thinking a food is either all good or all bad. Food marketers know that labels like "low-fat" and "all natural" unconsciously influence people to eat more, without noticing that the food that's low in fat might be high in sugar, for instance. Calling a food "healthy" often creates the expectation that it doesn't taste as good as the unhealthy alternative. Labeling a food as "bad" can also make it more attractive to kids by defining it as a special treat that's rarely available.

Automate Healthy Eating

Any of these techniques can be used not only to reduce mindless eating but also to bias unconscious habits away from candy and doughnuts and toward fruits, vegetables, nuts, and other healthy foods.

Because food quality matters more to health than food quantity, that approach may be more beneficial in the long run. For instance, try serving healthy foods in larger portions and making them more easily available than the junk food. Leave the salad bowl on the table and the steak and potatoes in the kitchen to bias second helpings in favor of salad. Make vegetables more appetizing by keeping dips around for snacking or by adding olive oil or cheese sauce during cooking. Contrary to common food anxieties, the vitamins and fiber do not disappear from vegetables if they're mixed with fat. If someone else in the house insists on keeping problem snack foods around, try to arrange for them to be stored in an inconvenient location.

In an environment designed to promote mindless eating, we have to work hard to avoid it. Unless eating is the main focus of a person's life, some decisions about what to eat are likely to be driven by habits. We would eat better and enjoy our food more if we automated the broccoli and paused to savor the ice cream, but many of us do it the other way around. Shoveling down chips in front of the television probably doesn't serve most people's food goals, unless one of them is to become as unhealthy as possible without enjoying the process. Instead keep fruits, vegetables, and nuts around for snacking while doing something else, and save the sugar and fat for times when they can be savored. The idea is not to become miserable with deprivation but to stay healthy and satisfied, eating as much as you want—and only you can decide how much that is—but no more.

For most people, experts recommend a goal of not gaining weight, which is achievable without heroic measures, rather than the often-doomed goal of losing a lot of weight and keeping it off. This chapter should provide some tools to help with that goal, as well as the bigger project of improving lifestyles to prevent or reverse some of the biggest killers in the modern world. People who focus on increasing physical activity, eating vegetables, and reducing stress are way ahead of most dieters when it comes to health. Gaining and losing the same twenty pounds over and over again is not an effective way to approach health concerns. If we commit to maintaining good habits most of the time, then the occasional indulgence (let's not call them slips) won't be more than a fond memory.

13

Healthy Is Better Than Thin

Emma Lewis dreads medical treatment because doctors tend to focus on her weight rather than her symptoms.[1] One evening she went out dancing for four hours and then sprained her ankle while walking home several miles in "slightly inadvisable shoes." When she consulted her doctor the next morning, he said she needed to get more exercise. That recommendation wasn't some new treatment for sprains. Instead he meant that it was more important for her to lose weight than to heal her damaged ankle. He didn't bother to find out that she already gets plenty of exercise and eats a whole-grain vegetarian diet, that she's a weight lifter who can squat 220 pounds, or that her blood pressure and blood sugar are normal. Instead he reasoned that because she's obese, she must be unhealthy, and losing weight should be her most important priority.

The assumption that body size directly determines the health of individuals is common, harmful, and wrong. In medicine, it leads to poor care and erodes trust between doctors and their patients, preventing some people from seeking medical care at all.[2] In society, it drives panic about the "obesity epidemic" and increases the pressure to diet, though dieting has never been shown to improve health in the long term.

Fortunately, opting out of our culture's obsession with weight does not mean we are fated to become sick and die young. Yes, obese people are more likely to be diagnosed with a variety of diseases and more likely to die prematurely than thinner people, but the evidence suggests that obesity is statistically associated with disease in large part because the factors that lead to diabetes and heart attacks also increase the odds of obesity. Those factors include lack of physical activity, stress, sleep loss, and eating a lot of processed foods. People can improve their health by changing the underlying behaviors—even if their weight remains the same—but instead many people and their doctors become fixated on body size as an end in itself. That's bad news, because of everything on that list, body size is the most difficult to modify. The good news is that body size is probably also the least important item for health.[3]

Unfortunately, most research on weight and health still focuses on weight loss. Funding is scarce for the longer-term studies that show diets to be ineffective or for interventions to prevent weight gain or to promote health without dieting. Only a few maverick scientists study how people can remain healthy at higher weights, a topic that deserves more attention. For example, Janet Tomiyama of the University of California, Los Angeles, runs the Health Registry of Obesity (HERO), a study aimed at figuring out which characteristics distinguish healthy obese people from their unhealthy peers.[4] By joining the registry, participants can help shift the medical conversation from weight loss to health.

Straight Talk on Weight and Risk

Throughout this book, I refer to people as "normal weight," "overweight," and "obese," but I want to clarify what I mean—and what I don't mean—by those terms. Public health officials define these categories based on body mass index (BMI), which is a person's weight in kilograms divided by the square of their height in meters. People with a BMI of 18.5 to 24.9 are in the normal-weight category. A BMI of

25 to 29.9 is defined as overweight, while anything over 30 is considered obese. Muscular athletes are often misclassified by this system (more than half the players in the National Football League in 2005 were technically obese),[5] but that's not a major issue, because most of us don't carry enough muscle to greatly increase our weight.

The big problem is that these category boundaries aren't the ones that make sense based on mortality risks, according to statistics from federal researchers at the Centers for Disease Control and Prevention.[6] Women at the low end of the so-called normal range, with a BMI of 18.5 to 20, have the same risk of early death as women at the low end of the obese range, between 30 and 35. Men have the same risks at the low end of the range defined as normal as they would at a BMI of 35 to 40. People in the so-called overweight range have the lowest risk, being a bit less likely to die prematurely than people in the normal weight range. Overweight people are more likely to be diagnosed with heart attacks or diabetes but are also less likely to die of these ailments than normal-weight people—a pattern called the obesity paradox, observed for a variety of health threats. Finally, the risk of premature death within the obese category doesn't increase much until people reach a BMI of 35. This evidence suggests that both the "normal" range and the "overweight" range are currently set too low.

The risks also change with age.[7] For people in their twenties and thirties, mortality risks stay low across a broad range of weights, from a BMI of 15 to a BMI of 30. As people age, weighing too little becomes more dangerous, while weighing too much becomes less dangerous. That probably happens because older people are more likely to become seriously ill, and carrying some extra weight provides a buffer against illness-associated weight loss. By age fifty, the low end of the normal range is risky. By age sixty-five, the health risks of obesity become balanced out by the benefits, leaving no statistical advantage to being in the normal range relative to the obese range. In particular, at that age, having a BMI of 19 carries the same mortality risks as a BMI of 40. For people over sixty—an important category for mortality statistics, as old age is when most people die—the least risky weight is in the so-called overweight range.

The weight ranges used to be defined more sensibly. In 1998, an expert panel convened by the National Institutes of Health lowered the definition of overweight from a BMI of 27.3 for women and 27.8 for men to 25 for everyone. The new definition caused 35.4 million people in the United States to become overweight overnight, without any change in their bodies.[8] Even at the time, most evidence suggested that the normal and overweight ranges were set too low. One possible explanation for the change is that seven of the nine panel members were the heads of diet clinics, which could arguably give them a conflict of interest in declaring tens of millions of people in need of their professional services.

Another potential conflict of interest is that the federal panel followed the lead of a World Health Organization report released in 1995, which was drafted by a task force primarily funded by companies that make weight-loss drugs, including Hoffmann–La Roche and Abbott Laboratories. Indeed, many obesity researchers have financial ties to drug companies or diet clinics. When we compare the research with their public health recommendations, it appears that those financial interests are having the effects that the companies intended. Although doctors often deny being influenced by gifts from drug companies, many studies show that their behavior reflects the funding, even if they aren't consciously aware of their biases. Considering that drug companies keep records on the prescribing behavior of individual doctors, we might wonder why those gifts would continue if they weren't effective.

Overweight and obesity in children are determined by a different method—several methods, in fact.[9] Children's weight varies depending on where they are in the growth cycle, so when they begin to put on weight, they may be getting ready to shoot up in height. In addition, most children are skinny until age six, when they begin to fill out. To deal with all this variability, public health officials typically define overweight as exceeding the 85th percentile of some reference BMI distribution for the child's age and sex, and obesity as exceeding the 95th percentile. That is, children who would have been within the top 15 percent of BMI for their age in an earlier era are considered overweight, and those who would have been in the top 5 percent are considered obese. A common reference population is taken from the National Health and Nutrition Survey II, done between 1976 and 1980. Thus, an eight-year-old boy in 2015 is labeled obese if 95 percent of males his age in the late 1970s had a lower BMI than he does. Using different reference populations or different percentile cutoffs can give vastly different estimates for the percentage of children who are overweight or obese. None of these estimates is directly related to health outcomes, as the potential increases in disease risk are still many decades away. That's why the decision to define a particular body size as abnormal in childhood must be somewhat arbitrary. It can also do a lot of harm: stigmatizing children for their body shape increases their risk of both eating disorders and weight gain. Given those risks, weight-loss expert Yoni Freedhoff argues that parents should not weigh their children at all.[10] Focusing on nutritious eating and fun physical activity as a family is a much better way to set children up for healthy lives than trying (and likely failing) to manage their weight.

Health, Not Thinness

Concentrating on healthy habits rather than weight is good advice for adults as well. Though most doctors recommend dieting to people above the normal weight range, that recommendation is not supported by evidence. No research has ever shown that losing weight through lifestyle changes improves health in people over the long run. That's true for one simple reason: no one knows how to get more than a few people to maintain a substantial weight loss for more than a couple of years. In laboratory animals, there is some evidence that permanent weight loss improves health. Unless you're willing to live in a cage with a food dispenser for the rest of your life, though, that research probably doesn't apply to you. What happens to people dealing with the options available in the real world should be our concern. The persistent difficulty of weight loss suggests a strategy that many experts now recommend: concentrate on maintaining a stable weight throughout adulthood. For people who eat well and exercise regularly, that weight doesn't need to be a particularly low one.

Even when health problems are directly caused by weight gain, that doesn't mean that weight loss is the only solution. For example, the hormone leptin, which is released by fat cells, acts directly on a brain region called the dorsomedial hypothalamus to increase blood pressure.[11] This outcome is not inevitable, though, because people can reduce their blood pressure through exercise, stress reduction, or medication. The only alternative for most people is to temporarily reduce their blood pressure by dieting and then watch it go back up when the weight returns.

A major effort to show that weight loss would improve health was the Look AHEAD study (the intensive weight-loss intervention with the underwhelming results), which was designed to determine whether long-term dieting would reduce heart disease in people with preexisting diabetes. After 9.6 years, the trial was stopped when a futility analysis showed that all that effort was unlikely to produce any difference between the intervention and control groups in cardiac health, even as more than half the participants maintained at least 5 percent weight loss—the modern medical definition of weight-loss success.[12] (Over time, that definition has gradually been scaled back from getting into the normal weight range to losing 20 percent of body weight to 10 percent and then to 5 percent, as studies accumulated to show how difficult it is to meet any of those goals. Nonetheless, weight-loss experts have insisted that each of these successively more modest goals has important health benefits.) An obvious argument is that losing 5.7 pounds is not enough to produce substantial health advantages, as it wouldn't move most people out of the overweight or obese categories. That response makes my point for me. If there is some amount of weight loss that would improve health, there is no known diet that can achieve it for most people—not even a diet with intensive long-term support in a highly motivated population.

A criticism of the epidemiology studies that fail to find reduced mortality associated with weight loss is that they are contaminated by people who have lost weight because they are sick, for example with cancer. To avoid this problem, some researchers refuse to consider data from any participants who might have lost weight due to illness by excluding, for instance, everyone in their sample who has ever smoked. The problem with that approach is that it requires throwing out the data from more than half of the population, including many smokers who don't have cancer—and who most likely differ in multiple ways from the participants who remain.[13] A more sensible way to evaluate the effects of intentional weight loss might be to ask people whether they are currently trying to lose weight, to avoid considering those who are losing weight accidentally due to illness, and then follow their health outcomes over future years. One such study found that people who lost weight between 1975 and 1981, after stating their intention to become thinner in 1975, were actually more likely to have died by 1993 than people who wanted to lose weight but instead remained stable or gained weight.[14]

Dieting Causes Weight Cycling

Successful long-term losers are hard to find, but people who repeatedly gain and lose weight are abundant. Whether we call it yo-yo dieting or weight cycling, this pattern is common. The increased likelihood of weight gain in weight cyclers is well established in the research literature.[15] Weight cycling may also cause physical health problems, though the evidence is mixed. Long-term studies find that people whose weight fluctuates are more likely to die during follow-up than those whose weight remains stable, regardless of their weight. This pattern was seen in Harvard alumni[16] (a population without many overweight people), middle-aged German men,[17] a representative sample of the U.S. population,[18] and the Framingham study of cardiac risk factors, among others.[19] But epidemiology involves a large number of statistical adjustments that can bias studies in various directions, so it's no surprise that other researchers disagree on such a contentious topic.

No association between weight cycling and mortality was found in two large studies that included measures of whether weight loss was intentional.[20,21] That additional information may account for the discrepant results, but other differences might be important, including differing ways of measuring weight cycling and adjustments for different potential confounding factors. Participants in one study were older than those in the studies that found an association, which makes sense as weight affects mortality less in older people. The other study found that all the increased mortality among weight cyclers was attributable to their weight gain during the study. Weight cyclers are likely to end up not just heavier but carrying more visceral fat—the most dangerous type for health—and less muscle than they had in the beginning. These changes might be the explanation for their poor health, rather than a statistical contaminant that should be factored out.

One possible explanation for these inconsistencies is that the health risks of weight cycling are especially severe in normal-weight people. Among U.S. dieters, 46 percent of women and 20 percent of men are within the normal weight range when they begin their diet, and weight cycling is likely to change their body composition.[22] In the study of the physiology of starvation mentioned earlier, men who lost 25 percent of their weight gained back more fat and less muscle during their recovery,[23] a pattern also observed in other people and in rats. The health damage due to weight cycling in the Framingham study was strongest in those who dieted as young adults, when people are least likely to be overweight. In Japanese men, a history of weight cycling was associated with poor glucose regulation, high triglycerides (the molecule into which fatty acids are packed for storage and transport), high blood pressure, and low HDL (good) cholesterol in normal-weight but not overweight people.[24,25] A small experimental study in five normal-weight women found that two cycles of losing and regaining ten pounds led to increases in triglycerides and blood pressure that lasted more than three months past the end of the diet.[26] The sure conclusion we can draw from all this work is that dieting does not improve health or longevity in people on average.

Rx: Fitness

Why would any doctor recommend an intervention with a high failure rate and no evidence for positive health effects in the long run? Part of the answer is that diets are often successful in the short term, which can make it easy to ignore their long-term weaknesses. But the more interesting issue is that people who go on diets typically do become healthier in the beginning. Part of the reason is that most attempts to lose weight involve increasing exercise and eating more vegetables, if only because of their low calorie density. Much research supports the idea that exercise and vegetables have a strong positive influence on health, but most dieters (and their doctors) attribute their improved health to the weight loss, which is usually temporary, rather than to the healthy habits that underlie it. That's a shame, because any health improvements that are due to weight loss are likely to be unsustainable. When dieters stop losing weight or start regaining, they often give up their healthy habits and lose their health gains. If the healthy habits were a goal in their own right, people wouldn't be tempted to drop them just because their weight goals turned out to be unreachable. Making any significant change in day-to-day life is hard, but we make things harder on ourselves by trying to make those changes while also fighting the brain's defended weight range. Instead it pays to focus on behaviors that are easier to improve, like exercise.

The first statistical link between physical activity and health was a study by a doctor named Jeremy Morris, who in 1949 began to compare London bus, tram, and trolley drivers with conductors.[27] The thirty-one thousand men came from similar backgrounds, but the drivers sat all day at work, while the conductors had to climb 500 to 750 steps a day collecting tickets. Morris found that conductors were 30 percent less likely than drivers to suffer a heart attack, and the ones they did have were less severe. He also showed that postal workers who delivered mail had 25 percent less risk of heart attacks than clerks who worked in an office at similarly prestigious jobs.

The argument about obesity versus fitness was already going on. To address it, a few years later Morris took advantage of the fact that bus workers wore uniforms supplied by their employer, Transport for London. When he compared the men's jacket and trouser sizes with their heart attack risk, he discovered that although the conductors were thinner than the drivers on average, even the heaviest conductors were still protected against heart attacks. Later studies by Morris's student Ralph Paffenbarger confirmed that fitness lowers heart attack risk, independent of body weight, in a wide variety of groups, from longshoremen working the docks of San Francisco to Harvard students training for professional careers. Both men were devoted exercisers. Morris, who had been active since childhood, lived to be ninety-nine, while Paffenbarger, whose research inspired him to start running at age forty-five, lived to eighty-four, despite a strong family history of heart disease.

One of Paffenbarger's best-known trainees is Steven Blair, an epidemiologist at the University of South Carolina. Growing up as a farm boy in Kansas, he was an unlikely prospect to become a famous scientist. After a concussion ended his college athletic career, he decided to go to graduate school to become a coach. The exercise physiology classes got him interested in the science of physical activity, but his real calling turned out to be in the epidemiology of how physical fitness influences health. He finds the evidence for the importance of fitness so convincing that he uses the bathroom in another building on campus to build more movement into his daily life.

Blair has repeatedly shown that no matter how much people weigh, more physical activity predicts better health. In a meta-analysis, he compared fit versus unfit people across weight categories.[28] People who exercise, even moderately, are 200 to 250 percent less likely to die prematurely than people who rarely get up from the couch, whether they fit into the normal-weight, overweight, or obese categories. In contrast, people who are both obese and fit have only 21 percent more risk than normal-weight, fit people. This much smaller effect was not statistically significant. (That is, statistical tests were not able to rule out the possibility that the finding was due to chance.) These data indicate that exercise habits are ten times more important than weight in determining the risk of early death.

Because obesity is correlated with low fitness, these results suggest that doctors have mistaken a correlated factor (obesity) for the causal factor (fitness) that explains most of the mortality risk. Among middle-class professionals, for example, cardiovascular fitness measured in a treadmill test is low in 9.2 percent of the normal-weight group, 19.4 percent of the overweight group, and 50.8 percent of the obese group.[29] Blair estimates that low fitness is responsible for 16 to 17 percent of deaths in the United States, while obesity accounts for only 2 to 3 percent once the effects of fitness are factored out.[30] His research also suggests that exercise has a stronger effect on health for obese people than it does for normal-weight people. That makes it particularly important for doctors to encourage their obese patients to exercise.

Physical activity also matters for preventing and treating diabetes. Exercising muscles take up glucose for energy, so exercise reduces the amount of glucose in the blood. Researchers in China compared diet alone, exercise alone, and diet plus exercise for preventing people with impaired glucose tolerance from developing diabetes.[31] Over the six-year follow-up, exercise had the strongest effect, reducing the onset of diabetes by 46 percent compared with no intervention. The diet-plus-exercise group experienced the same effect as those who just exercised. A review by the Cochrane Collaboration, recognized experts in evaluating medical evidence, concluded "that exercise improves blood sugar control and that this effect is evident even without weight loss" in people who already have diabetes.[32] Another Cochrane review concludes more broadly that "exercise improves health even if no weight is lost."[33]

Maybe instead of relying on weight to diagnose health, doctors could use health markers, like blood pressure, cholesterol, triglycerides, and glucose tolerance. Another meta-analysis suggested that approach would be much more accurate.[34] Obese people with health markers in the normal range were only 19 percent more likely to die during the study than normal-weight people with good numbers. As in the fitness meta-analysis, the outcomes for normal-weight people with bad health numbers were more than ten times worse, with more than a threefold increased risk. Both overweight and obese people with bad health numbers fared slightly better than their normal-weight counterparts (though not well). In sum, weight carries little power to predict health outcomes beyond the extent to which it predicts fitness and good health markers. Both are more common among normal-weight people, but they don't need to be.

Doctors who prescribe weight loss to patients who aren't exercising regularly are choosing to promote an intervention with a high failure rate, instead of a lifestyle change with much lower barriers to entry and well-documented benefits. The focus on weight encourages us to abandon habits that are working to improve our health when we stop losing weight or start to regain it. Exercise, good food choices, and stress reduction could go a long way toward making all of us healthier without relying on the mirage of weight loss.

How Exercise Protects Health

An underlying cellular process links obesity, fitness, and stress to health problems like diabetes and heart disease. Cellular inflammation is a response to metabolic disturbances that can end up hurting the body. As we learned earlier, inflammation in the brain may cause the defended weight range to increase. One way to produce this inflammatory response is for a fat cell to exceed its storage capacity.

Fat itself is not the problem, as long as it's tucked away, but fat that isn't stored properly causes an inflammatory response. Cellular inflammation occurs progressively more as body weight increases, not only in fat cells but also in the liver, pancreas, and brain.[35] We can tell that fat storage matters more to health than fat abundance from two examples. First, removing more than twenty pounds of abdominal fat from obese people by liposuction does not improve their insulin sensitivity or risk factors for heart disease because liposuction doesn't reduce the amount of fat stored in each of the remaining fat cells.[36] Weight lost by liposuction is typically regained elsewhere on the body within a few months, as we would expect due to the weight thermostat. Second, people and laboratory animals with a genetic condition that keeps them from producing any fat-storage cells are very thin but also susceptible to many obesity-associated diseases. They have high blood triglycerides and severe insulin resistance. Because their fat cannot accumulate in fat cells where it belongs, it ends up in the liver (where it increases the production of glucose, triglycerides, and cholesterol, eventually causing cirrhosis), in the insulin-producing cells (where it leads to diabetes), and in the heart (where it promotes heart disease).

Fat tissue can expand by increasing the number of fat-storage cells, which produces little inflammation, or by increasing the size of individual cells, which causes inflammation.[37] One reason improperly stored fat is a problem is that saturated fatty acids can directly activate receptors that trigger an immune response. Larger fat cells are less sensitive to insulin than smaller ones, which may contribute to the development of diabetes. In adults, most fat is added by packing it into existing cells because new fat cells are produced only to replace old ones. Treating people with a drug that promotes the addition of new fat cells reduces the amount of fat stored in individual cells and improves insulin sensitivity.[38] Similarly, when mice that had become obese and diabetic by eating a high-fat diet were injected with compounds that stimulate fat-cell formation, their insulin sensitivity improved, and less fat was stored in their livers.[39] Unfortunately, these drugs that improve the health of fat cells also increase the risk of cancer, so they cannot be used routinely in people. On the other hand, cellular inflammation can be reduced in a variety of ways that don't involve weight loss, including through exercise.

Inflammation is also increased by factors other than obesity. For example, loneliness increases the risk of early death by 45 percent.[40] Social isolation is epidemic in the modern world. A quarter of adults in the United States know no one with whom they can discuss important matters. Loneliness increases inflammation, reduces sleep quality, and increases the release of stress hormones.[41,42] Socially isolated mice also show increased inflammation, which makes them more vulnerable to neural injury.[43] All these mechanisms may contribute to the increased risk of metabolic syndrome in lonely people.

Not all fat cells are created equal, as we learned earlier. Visceral fat, stored inside the abdomen, is dangerous because it releases fatty acids easily, allowing them into the blood where they can cause inflammation.[44] Because visceral fat is more metabolically active, it is preferentially lost during exercise, with or without weight loss, while fat beneath the skin is more resistant to change. Metabolically healthy obese people have little visceral fat and exercise often. Among obese men of the same weight, those with little visceral fat tend to resemble thin people in their metabolic profiles. Those with a lot of visceral fat, on the other hand, typically have insulin resistance and high triglycerides, early predictors of problems like diabetes and heart disease.[45] This helps to explain why fitness has a stronger effect than body size on health.

If inflammation is the common underlying cause behind diseases linked to metabolic disturbances, anti-inflammatory drugs might be useful in preventing these health problems. Aspirin and statins, which are both anti-inflammatory, do help to keep metabolic syndrome from progressing to heart disease or diabetes.[46] Indeed, some researchers believe that the cholesterol-lowering effects of statins are just a side effect of their main action of reducing inflammation. Doctors are testing a variety of other drugs that may block inflammation and thus break the link between lifestyle issues—like poor diet, lack of exercise, and stress—and their health consequences.

Another approach that may protect against inflammation is extended overnight fasting. Artificial lighting and easily available food now allow people to eat all day long and then snack well into the evening. Evidence from laboratory animals suggests that this habit harms metabolism. In mice, when food is available for only twelve hours during the waking period each day, high-fat and high-sugar diets cause less weight gain and less fat storage, without any change in total calories eaten.[47] Time-limited feeding decreases inflammation, improves blood cholesterol and triglycerides, and protects against insulin resistance. The same benefits result from twelve-hour feeding on weekdays with twenty-four-hour access to food on weekends.

Whether these findings apply to people remains to be seen, as few randomized clinical trials have been done. In one trial, normal-weight adults showed little difference in body composition or health markers if they ate one meal (fasting twenty hours) or three meals (fasting twelve hours) per day,[48] but the study didn't look at whether longer daily periods of eating might be less healthy, as seen in mice. Time-restricted feeding in overweight and obese people has been tested only for weight loss, so the studies done so far include calorie restriction. Under those conditions, people show better insulin sensitivity, lower triglycerides, and less hunger for the same amount of food if they eat earlier in the day.[49] The animal research suggests that time-limited feeding may reduce metabolic damage resulting from an unhealthy diet for people with limited options, but avoiding such damage in the first place might be a better strategy.

Healthy Habits Matter More Than Weight

Lifestyle choices contribute to most of the medical problems that kill us these days. In my TED talk, I presented the results from a study showing that four health habits predict much of the risk of dying over the next fourteen years, regardless of weight.[50] The habits were not smoking, exercising at least twelve times a month, eating five servings of fruits and vegetables a day, and light to moderate drinking (more than zero but less than one drink per day for women, two for men). Among people with none of these habits, the obese group had almost seven times more risk than people in the normal weight range, and overweight people had almost four times the risk. But for people with healthier lifestyles, weight mattered considerably less, with similar risks in all weight categories for those who practiced two, three, or four healthy habits.

Another study, of Europeans in their seventies and eighties, defined healthy habits as exercising, drinking moderately, not smoking, and eating a Mediterranean diet, which includes lots of vegetables and fruits, fish, nuts, and healthy fats like olive oil, with limited red meat and dairy products.[51] Over ten years of follow-up, people who exercised or didn't smoke were two thirds as likely to die as those who didn't exercise or did smoke. The Mediterranean diet or moderate alcohol groups had about three quarters of the normal risk. Each additional healthy habit increased survival. People with all four healthy habits were only one third as likely to die during the study as people with one or none. The authors estimated that if everyone adhered to these four habits, 60 percent of the total deaths during the study could have been avoided, even though about 60 percent of the participants were overweight or obese.

If lifestyle changes are important for health, while weight loss is merely a side effect, then we would expect the health benefits of dieting not to depend on the amount of weight that dieters lose. Indeed, that's what researchers find. A review article analyzed twenty-one studies that compared the health effects of weight-loss diets versus not dieting, in trials that randomly assigned participants to groups.[52] On average, dieters had maintained a weight loss of 2.1 pounds more than nondieters by two years after the diet began. At that time, the diets had no effect on cholesterol or triglycerides and had reduced blood pressure only slightly. Two of the reviewed studies found that dieting reduced the diagnosis of new cases of diabetes by 58 percent, but both of those diet plans included substantial exercise prescriptions, which received good compliance from the participants and probably explain the protective effect. Across all the studies, the amount of weight lost was not correlated with any of these health improvements, suggesting that eating and exercise changes may have been responsible.

To find out if good health habits are what really matter, researchers have tested interventions that emphasize lifestyle improvements without a goal of weight loss, an approach that nutrition professor Linda Bacon named "Health at Every Size." Her program promotes body acceptance, eating in response to hunger and fullness, making healthy food choices, moving the body in ways that feel good, and seeking out social support. In one study, overweight and obese women were randomly assigned to a weight-loss diet or to the Health at Every Size program.[53] The program started with six months of weekly sessions on each of those topics, followed by six months of monthly group meetings for continuing support. The nondieting group lowered their total cholesterol, LDL (bad) cholesterol, systolic blood pressure, and triglycerides, and those improvements lasted for a year after the program ended. (That's two years after it started, according to how diet results are usually reported.) Measures of disordered eating also declined in this group. The matched group of dieters lost weight and showed health improvements at the end of the intervention year, but by the next year they had regained the weight, and all their health improvements had reversed. A similar program in New Zealand also significantly improved blood pressure at a two-year follow-up.[54] In these programs and several others, nondiet participants improved their self-esteem and decreased their body dissatisfaction as well.

People who practice Health at Every Size, for themselves or their patients, believe that we need to declare a truce in the war on obesity. Fighting against fat puts people into conflict with their brain's energy-balance system. It has only one job, keeping weight within the defended range, which it does all the time without distraction. Good nutrition and exercise can improve health and avoid the toxic side effects of dieting, which are goals worth striving toward. Given that most of us have responsibilities beyond keeping our weight down, the war on obesity is simply not a fair fight. Let's make peace with food and move on with our lives.

14

Change Your Lifestyle, Change Your Health

Personal trainer Kelly Coffey loves the clients who come to her because they want to improve their fitness without worrying about their weight. "Increasing your cardiovascular capacity, getting stronger, anyone can do that," she says. "It doesn't matter how old you are. It doesn't matter how fat or thin you are. Getting more fit is the easiest goal I could possibly hope to help someone meet." To improve health and well-being, it's also arguably the most important. When we think of exercise or eating well as a means to achieve weight loss—or worse yet, as punishment for weight gain—rather than important goals in their own right, we often end up sabotaging our efforts to create a healthier lifestyle.

The Joy of Exercise

Increasing physical activity improves many aspects of health all by itself. A federal government report concludes that exercise definitely reduces the likelihood of early death, heart disease, stroke, high blood pressure, type 2 diabetes, breast and colon cancer, weight gain, falls resulting in injury among the elderly, depression, and age-related cognitive decline, including Alzheimer's disease.[1] In addition, some evidence suggests that it strengthens bones. A meta-analysis involving more than 100,000 people concluded that highly fit people are 42 percent less likely to die prematurely than sedentary people.[2] Specifically, each increase in fitness that improves jogging or running speed by one mile per hour is associated with a 22.6 percent reduction in the risk of early death. On average, people who exercise twice a week starting in middle age are half as likely to be diagnosed with dementia in old age.[3] New converts may also notice immediate benefits to exercise, like improved sleep quality, increased strength and endurance, and stress relief.

A modest amount of exercise may be necessary for the brain's energy-balance system to work correctly.[4] In the 1950s, Jean Mayer studied diet and exercise in jute mill workers of Bengal. Among workers whose job required movement, even as little effort as a mechanic walking around the shop, calorie intake balanced exactly with physical exertion. In contrast, office workers ate more than their energy expenditure required, and the most sedentary workers ate more than their slightly less sedentary colleagues. These results may explain why people who don't exercise are vulnerable to the slow weight gain of one or two pounds a year that often occurs with age. On average, an increase in exercise doesn't lead to weight loss because the brain's energy-balance system increases food intake to compensate, but there's a lot of individual variability in that compensation, so this approach does work for a few people.

People who've used exercise only for weight control are in for a pleasant surprise when they try physical activity for its own sake. Moving the body when it isn't starving is a totally different experience. When people are below their defended weight range, the brain's energy-balance system is trying to conserve energy by reducing their motivation to be active and by forcing their muscles to do more work with less energy. Those changes make people feel tired and sluggish during physical activity because the brain wants them to sit down and conserve energy. In contrast, a well-fed body is much more likely to feel energetic. Distance walking is what our bodies are adapted to do by evolution, but they need fuel like any other machine.

The heavier people are, the more health benefits they're likely to gain from becoming fit. Normal-weight people with low fitness are 2.2 times more likely to die over a ten-year follow-up than their fit counterparts. Obese people with low fitness are 3.1 times more likely to die compared with the same reference group, while fit obese people have only 1.1 times the risk.[5] These statistics support the idea that fitness makes a much larger difference to mortality in obese people. Exercise fights inflammation, so it can also help to break the link between stored fat and metabolic diseases.

Starting to exercise can present practical and psychological difficulties for people who are obese. Finding a fat-friendly class or facility can help. Most exercises can be modified to accommodate people of any fitness level or body shape, and there are approaches designed for people of all weights. In addition, others who have succeeded are available to answer questions and provide advice, on Web sites like the Fit Fatties Forum[6] and in books like *The Unapologetic Fat Girl's Guide to Exercise and Other Incendiary Acts* by Hanne Blank.

The biggest health improvements occur when people first get up off the couch. Someone who isn't physically active at all can expect to gain another three to five years of life and better health just by walking for 150 minutes a week. As little as 10 minutes a day helps, and the benefits continue to increase at least up to an hour a day of moderate to vigorous exercise. Interval training, in which exercisers alternate brief periods of high effort (thirty seconds to three minutes) with longer periods of mild or no exercise (one to five minutes), produces more fitness gains than continuous moderate activity for the same average amount of energy expenditure.[7] A person who's never been active before might start by walking for five minutes every day, then build up abilities over time. After deciding to make exercise a permanent part of life, people have plenty of time to implement that decision gently and are more likely to stick with the plan if they take it slowly in the beginning and don't hurt themselves. Although I had tried and failed to exercise regularly many times before, I found that when I quit dieting, I was able to use the willpower that I had freed up to finally succeed in forming a daily activity habit.

Even if they exercise regularly, people who spend long hours sitting without interruption have an increased risk of death, no matter how much they weigh. A meta-analysis of studies involving almost 600,000 people estimated that sitting for ten hours per day increased the risk of early death by 34 percent after physical activity was taken into account.[8] That result suggests that 5.9 percent of total deaths in the United States are attributable to sitting, independent of exercise. For comparison, 5.1 percent of deaths are due to stroke. Less than three hours of sitting per day posed no extra risk, and the strongest effects occurred in people who sat more than seven hours a day. To avoid this risk, anyone who has a desk job or spends long hours in front of a screen at home should take a break and stand up for a few minutes two or three times an hour.

Prolonged sitting is associated with high triglycerides, insulin, and waist circumference, and with low HDL (good) cholesterol.[9] In one study, overweight middle-aged adults were asked to take a two-minute break to walk at two miles per hour after every twenty minutes of sitting or to sit for five hours without getting up.[10] Then they were given a test meal containing fat and sugar. The participants who took breaks showed less insulin release and lower blood sugar than the long-term sitters.

For readers who are currently sedentary and make only one change as a result of reading this book, I hope that it's deciding to get some exercise. Take a walk every day or go for a run or a bike ride or a swim a few times a week. Park at the far end of the parking lot or get off the bus or subway a stop earlier than your destination. Try yoga, in a class with an instructor or by video in the privacy of your home. Go dancing or take the grandchildren to the park—anything that's enough fun to do regularly. For people who can't think of anything active that they enjoy doing, try making a commitment to sample a new activity every week or two for a year. Ability doesn't matter as much as interest, sticking with it, and slow improvement.

The best time to develop a habit of physical activity is during childhood. Sedentary children have a high risk of growing up to be sedentary adults. Less intuitively, a lot of active children stop being active in their early teens, due to self-consciousness or lack of time, and most of them remain sedentary as adults too. Getting young kids to run around is usually not difficult. Parents can direct that energy toward the development of skills and habits that will be useful in adulthood. By their twenties, most people have stopped climbing trees and jumping rope, but a love of softball, swimming, or bicycling could be a lifelong affair. It's important to encourage teenagers to stay involved in physical activities they enjoy. That might mean taking them and some friends hiking or paying for dance classes. Watch out for activities that promote weight loss or body dissatisfaction. One of my friends took her four-year-old daughter out of ballet class because the teachers often discussed weight and switched her to Bollywood dancing, which accepts a variety of body shapes.

Starting an exercise habit as an adult requires commitment, but not necessarily the joyless variety. One approach is to find a child or a dog who enjoys being an activity partner. Especially in the early stages of habit formation, having an enthusiastic buddy can make the difference in getting out the door consistently. An unenthusiastic partner can have the opposite effect, so partnering with a friend is a good idea only if both are equally committed. Elderly people walked less often if they were paired with a walking partner from the retirement home, because if one partner didn't want to go out, both stayed home. In contrast, participants walked more often if they were asked to walk a dog. People who like dogs but don't have one of their own might consider volunteering to walk the dogs at the local shelter or finding out if a neighbor's dog would appreciate an extra walk. Getting the habit started is the hardest part for most people. College students without an exercise habit who were paid $125 to go to the gym nine times in a month were likely to continue to go after the study ended, with about a third of them visiting the gym once a week.[11]

Eat for Health

Eating a better diet can also lead to major health improvements, but there's a lot of confusion over what constitutes healthy eating. Despite the controversies, nutrition research does converge on some clear recommendations.[12] The marketers of processed food deliberately create some of that confusion, but public health officials also bear part of the responsibility as unintended consequences of their recommendations in the 1970s and 1980s. Because of epidemiology research linking heart disease risk to consumption of saturated fats, found in red meat and dairy, doctors started pushing people to change their eating habits—just at the moment in history when regular Americans were getting rich enough to eat steak regularly and when the processed-food industry was growing. In addition to recommending that we reduce saturated fats, health officials pushed to reduce all fats to less than 20 percent of total calories, even though many populations of healthy people get a much higher percentage of their calories from fat.

The problem, common in dietary interventions, is that when certain foods are cut out of the diet, they are replaced by other foods, which are often not the ones health experts would have chosen. Habits largely control eating, and when people are trying to change old habits, they become especially vulnerable to outside influences. In this case, marketers of processed foods took advantage of the opening that public health officials provided. Across the population, people trying to reduce their fat intake replaced fats with refined carbohydrates, as food companies positioned products like SnackWell's cookies as low-fat, "healthy" alternatives. Though popular with consumers eager to eat cookies without guilt, this dietary change was not a good idea. Not only are white flour and added sugars almost certainly worse for health than saturated fats, but also as consumers switched from butter to margarine, they started eating trans fats, which increase heart disease risks even more than the saturated fats they replaced. To be fair, no one in authority said, "For your health, you need to start eating more processed foods" (though they did suggest switching to margarine—oops), but that's what happened. In the end, people ate the same amount of fat while adding more carbohydrate calories to their diet. That change reduced the percentage of calories from fat in the American diet, but not in a good way.

Of the important healthy habits, eating well may be the most difficult for veterans of the diet wars, which have produced a ridiculous amount of fierce talk about "the only right way to eat." The truth is that there is no One True Way, but we certainly have discovered some wrong ways to eat over the past half century or so. People have been feeding themselves for hundreds of thousands of years, in places that were much more challenging to navigate than the supermarket. Unfortunately, the skills and habits that got our ancestors through those tough times are working against many of us today. When food is scarce, it makes sense to conserve energy whenever possible and feast when the opportunity presents itself, but those same tendencies can cause problems when food is abundant, high in calories, and low in nutrients.

There's no point in diving into the low-fat versus low-carb weight-loss wars, because we're talking about health here. All we need to know about those zones of diet enthusiasm is that neither one changes the defended range of the energy-balance system. That's why both approaches are equally unsuccessful in the long term, producing similar short-term losses that are regained over two to five years.

Most experts agree that the replacement of fats with refined carbohydrates was a net loss for health. Refined carbohydrates interfere with the job of the brain's energy-balance system, as the satiety that people experience after eating them does not appear to be appropriate for the calories they contain. That fact alone could explain why the increase in refined carbohydrates in the food supply coincided with the increase in obesity, and it argues for increasing consumption of whole grains rather than treating all carbohydrates as suspect.

We often get hung up on the difficulty of making major changes to our diets—should we become vegan or give up carbohydrates?—when most of us could improve our health with a few changes that almost every expert would consider beneficial. Most people notice these changes, so they require some effort, but at least they don't require us to pick a fight with our brains that we'll probably lose. Focusing on perfection is an excellent way to become paralyzed, so aim to make eating habits better rather than flawless. Just expanding dietary horizons can help, because people are likely to be healthier if they eat a wide variety of foods than if they eat the same things all the time. Here are two uncontroversial dietary goals whose health benefits are supported by considerable evidence.

First, eat more vegetables, because they contain fiber and many nutrients that aren't available in a vitamin pill. There is no disagreement on this recommendation, no matter what nutrition expert or plan you're following. Most of us don't get anywhere close to the recommended five to thirteen servings of fruits and vegetables a day. A serving is one half cup for most vegetables, and every serving helps. To put it more intuitively, at least half the food on the plate should be vegetables and fruits. The U.S. Centers for Disease Control and Prevention estimates that the average person in the United States eats vegetables 1.6 times per day, and 22.6 percent of the population doesn't manage to encounter them even once a day.[13] For adolescents, more than a third (37.7 percent) don't eat vegetables once a day. If fresh vegetables are too much trouble, frozen ones have plenty of nutrition, as do precut salads. If someone in the household doesn't like vegetables, try throwing some greens into soups or stews, or blending larger vegetables as needed to keep the kids from picking them out. Also try some of the tricks for converting vegetable haters into vegetable eaters mentioned earlier. Compared with people who ate seven or more servings of vegetables per day, people who ate them less than once a day were 50 percent more likely to die over a seven-year follow-up in one English study.[14] Fruit was less strongly associated with mortality, but it still helps, and many people like fruits better than they like vegetables.

Second, minimize added sugars, refined grains like white flour, and the processed foods that contain them. This change will reduce the risk of metabolic disease, especially diabetes, and help the brain's energy-balance system work correctly, as it hasn't had much experience with foods that were made in a factory. A major dietary change over the past few decades is the increasing consumption of convenience foods, both take-out and prepared meals from the grocery store. Added sugars show up where we wouldn't expect them, in foods like spaghetti sauce and cereals. Read nutrition labels to find the exceptions or make food at home from whole ingredients.

Instead eat foods that contain dietary fiber, including oatmeal, nuts, beans and peas, and whole wheat, in addition to vegetables and fruits. Fiber is important for the health of gut bacteria, particularly the ones that fight inflammation and help calibrate the developing immune system to prevent autoimmune diseases. A diet high in fiber lowers cholesterol as much as statin drugs, and four times as much as a low-fat diet.[15] A high-fiber diet also improves blood-sugar control in diabetics, probably because it moderates the spike in blood sugar induced by eating.[16] The 2010 Dietary Guidelines recommend twenty-five grams of fiber per day for women and thirty-eight grams for men, an amount that is nearly impossible to get from processed foods. Be careful to avoid processed foods that include small amounts of fiber to create a "health halo" around products that are not actually nutritious.

Many people feel that they're too busy to get along without processed foods (an attitude that the companies have worked hard to encourage), but the time savings from using them are mostly an illusion. Families who serve frozen food for dinner save only ten to twelve minutes, on average, compared with nights when they cook from scratch.[17] The health problems, unfortunately, are real. People who can find five or ten simple recipes that their family likes, or take a couple of hours on the weekend to cook a big batch of something to freeze, can improve the health of the household without spending much extra time. In general, try to eat food made from fresh ingredients more often and food found in boxes and bags less often.

Don't worry about cutting back on the total amount of fat in the diet. The fats found in avocados, fish, nuts, and vegetable oils are good for us. The role of saturated fats, the type found in dairy products and red meat, in a healthy diet is more controversial. There's no question that replacing saturated fats with refined carbohydrates is a net loss, but replacing saturated fats with healthier fats does improve health modestly. A meta-analysis found that replacing saturated fats with polyunsaturated fats reduces the risk of heart disease by 19 percent,[18] and a similar analysis from the Cochrane group found a 14 percent reduction.[19] On the other hand, neither study found a decrease in deaths, which is the most important outcome for health epidemiology.

The 2015 Dietary Guidelines for Americans declares that we can stop worrying about how much cholesterol we eat. For most people, high-cholesterol foods like eggs and shrimp have no influence on blood cholesterol. For those whose cholesterol does respond to diet, eating eggs raises both HDL (good) and LDL (bad) cholesterol in the blood.[20] Taken together, this research suggests that concerns about the types of fat in the diet should not be a priority for most people. We can gain much more influence over our health by increasing exercise and vegetables and decreasing white flours and sugars.

One tasty option is the Mediterranean diet—a way of eating, not a plan for weight loss—which centers around olive oil, fish, nuts, grains, beans, vegetables, and fruits, with smaller servings of red meat and dairy. Because this diet gets 35 to 40 percent of its calories from fat, most people find it easy to maintain, which matters a lot to someone who's trying to sell a lifestyle change to kids or a grumpy spouse. Its advantages for heart health are at least as strong as those of a low-fat vegetarian diet,[21] and in intervention studies it improves metabolic syndrome symptoms in obese people.[22] This delicious way of eating blows away the image of healthy living as a joyless march to the grave while nibbling on twigs and leaves (another unintended consequence of three decades of low-fat public health advice). There's only one catch: at least one person in your household needs to be willing to cook.

Without expert label reading, which is not a fun hobby, the easiest way to eat with fresh ingredients is to cook. In two generations, cooking has gone from something that happened almost every night in almost every house to a spectator sport on TV. Reversing that change is going to be an effort, but it's hard to avoid if we want healthy food.

The good news is that regular daily cooking bears little resemblance to the feats on cooking shows. Baking a whole chicken hardly requires more effort than taking the plastic off, pulling out the bag of giblets (do not forget this step), and sticking it in the oven. Toss in a few baking potatoes or sweet potatoes with the chicken, thaw a bag of frozen vegetables, and the meal requires less hands-on effort than going through the drive-through of McDonald's.

To get started, buy a general cookbook like Mark Bittman's *How to Cook Everything* or, for people with a lot of time pressure, *How to Cook Everything Fast*. Both books are written for people who don't have much experience with cooking. Try making dinner on a weekend when there's some extra time available, and gradually expand the effort until someone is cooking most nights. By letting kids help with cooking, parents are not only preparing the kids to eat well in adulthood but also setting up the family to have competent help in the kitchen on nights when parents are too busy to cook. Even if parents are learning themselves, children are likely to enjoy learning at the same time.

My favorite benefit of cooking is that it takes care of most nutritional concerns without additional thought. The brain's energy-balance system can deal accurately with all the dishes that are prepared at home, so people can eat when they're hungry and stop when they're full without being concerned about weight gain. The food is not designed to create cravings that lead to overeating. Because people have been eating whole foods for millennia, there's no need to think about unexpected health hazards that may arise when we experiment on large populations with industrially created food products. In the long run, this approach is likely to lead to a relationship with food that involves pleasure rather than counting, weighing, or worrying but still leaves everyone well nourished.

"When I was chronically dieting and so preoccupied with all the bad stuff going on in my life, I was totally living in my head. I wasn't even thinking about how my body felt," recalls Anne Golob. She'd been fighting to lose twenty pounds for decades, and she was stressed and miserable when she first walked into Kelly Coffey's studio. Now that Anne has gotten back in touch with her body's signals, she notices how much better she feels when she's sleeping well, eating good food, meditating, and exercising regularly. She's more alert and less grumpy during the day, and the arthritis pain in her shoulder bothers her less than it used to. This practice has also made her more accepting of her body: its shape, its abilities, and its needs.

The transition from yo-yo dieting to focusing on health can be difficult. Many veterans of the diet wars end up with persistent food anxieties that are made worse by post-traumatic reactions to health advice. Plans for long-term deprivation with the goal of weight loss are often described as "not a diet, but a lifestyle change," which makes it harder to think about lifestyle changes in a positive way. After multiple attempts to deprive ourselves in exactly the right way, in pursuit of a payoff that never stuck around long, if it arrived at all, we become conditioned to think of food as a battleground and exercise as a punishment. Both these attitudes work against successful behavior change. Sometimes they paralyze us so thoroughly that we can't figure out what to have for dinner at all. That's not a recipe for healthy eating. Giving ourselves permission to eat when we're hungry is the first step, but it may not be enough. Before we can respond effectively to internal cues of hunger and fullness, we may need to learn to accept that we can be good people without starving ourselves.

The key to long-term success is to make lifestyle improvements slow and positive. Try to add or substitute habits rather than forbidding or taking away foods. People who don't enjoy their food or hate going to the gym are unlikely to stick with the program long enough to see the benefits. For all these reasons, changes that don't drive anyone in the house crazy are likely to work out better in the long run. Start small, by replacing one unhealthy snack with nuts or fruit, or by walking around the block every night after work. After getting used to one new habit, when it has become automatic, make another improvement. Remember that after giving up dieting, we no longer need to factor in time spent going in the wrong direction, so we can afford to take a slow and steady attitude to behavioral change by adopting one beneficial habit at a time.

15

Good Habits Beat Good Intentions

When I have time, I like to catch up on the work of other science writers. I can spend hours at my computer learning about new ideas and absorbing tricks of the trade. My Twitter feed always turns up something interesting—which recently became a serious problem. When I first sat down to write this book, I kept finding myself on social media, reading articles in spite of my clear intention to spend the day writing about a topic that I find engaging. Eventually I realized that sitting at the computer was the cue that triggered a habit of Web surfing. Without conscious intent and with barely any awareness, I would click a link on the browser, then come out of a daze when half the morning was gone. That habit turned out to be so strong that I had to install blocking software to keep me off the sites that give me trouble during work hours. Now when I sit at the computer, the first thing I do is open my manuscript, whether I plan to work on it or not. Through repetition, I've substituted a new habit for the old one.

As the fossilized relics of our past experiences, habits are where good intentions go to die. Through habits, actions that once led to rewards continue to drive our behavior, even when those behaviors no longer serve our goals. Automating routine actions frees up brain space for other projects, but at the cost of limiting flexibility. The stronger people's habits become, the less their intentions determine what they will do in a given situation.[1] Most people are not aware of this effect, though, as those with the strongest habits tend to express the most confidence that their intentions predict their future behavior.

Curiously, in light of these findings, most programs designed to promote behavioral change focus on motivating people to form appropriate intentions rather than teaching them how to change their habits. Less surprisingly, the results of such programs are often disappointing, as the new intentions fail to influence behavior.[2] The expectation that we can use willpower to force ourselves to behave differently over the long term is not supported by research, as I've said before. Most of us don't have enough willpower to spare. Besides, declaring war on the unconscious brain is doomed to failure because the unconscious mind is on duty all the time, while willpower takes a lot of vacations.[3]

Instead of fighting with our brains, we can accept their limitations and arrange the circumstances to create new habits that support our well-being. Starting or changing habits requires both willpower, to help us continue through the inevitable struggles, and mindfulness, to keep us aware of our actions as they occur. That's one reason to make such efforts during low-stress times in our lives, when we have some spare mental capacity to improve the odds of success.

How to Change a Habit

As well-trodden paths in our brains, habits take a long time to learn and a long time to change. They're established by repetition of a sequence in which a cue leads to a behavior that leads to a reward. Indulging a habit doesn't require attention, but resisting one does. People (and other animals) are likely to develop habits in situations where the same cue triggers the same response repeatedly, like brushing our teeth after dinner, which is rewarding because it makes them feel clean. Actions that need to be flexibly adjusted as conditions change usually do not become habits. To avoid making a habit out of the drive to work, we could take a different route each day. People who live alone report more habitual behavior than people who live with others. Daily actions are almost certainly habits, weekly actions are likely to be habits, and actions that happen once a month or less often are probably not habits. Not all habits are actions: some are thoughts that are automatically triggered in a certain context. Such thoughts are an important component of anxiety and depression, and a key aspect of cognitive-behavioral therapy is becoming aware of them and challenging their content. The more we practice our habits, the more they determine our behavior, in part because habits suppress the brain circuits that control competing behaviors.[4]

The defining feature of a habit is that the behavior continues when the reward that established it is removed. In the laboratory, scientists test whether a behavior is habitual or goal directed by making the reward less valuable. For example, they might add bitter taste to a water reward. If animals do a task to achieve a reward, they will work less hard for a smaller reward and stop working if the reward loses its attraction. If they can get the same reward another way, with less effort, they will switch to the new task. Animals that are doing the task out of habit look exactly the same when the normal reward is effective, but they act differently from goal-oriented animals when it's no longer attractive. After they've eaten so much food that they no longer want more, for example, habit-driven animals continue to repeat the original task in an automatic way. They also don't switch strategies when the original way of doing things stops working.

Most of us have some behaviors that were rewarding initially, but have become so habitual that we keep doing the same thing long after it has stopped serving our goals. Perhaps, like those laboratory rats, we continue to eat food we don't want, or we spend too much time on the couch even though it makes us sluggish and uncomfortable. Willpower may suppress those habits for a while. Eventually, though, when we get busy or overwhelmed, the unwanted habits will return unless we have replaced them with new ones.

Habits are learned in different parts of the brain from goal-directed behavior. People with damage to an area called the dorsal striatum have normal conscious learning but fail to learn procedures and habits. In laboratory animals, damage to one part of the dorsal striatum prevents habit formation no matter how long the animal is trained, while damage to another part causes behaviors to become habitual after less practice than normal. Similar effects are found after damage to two nearby parts of the prefrontal cortex, one that supports goal-directed behavior and another that drives habit development. These circuits are important for everyday habits, and their malfunction can lead to disorders like addiction and obsessive-compulsive disorder.

The opposite of habitual behavior is mindful behavior. Mindfulness enthusiasts encourage students to try to be aware of each moment of their lives, but being mindful doesn't require using deliberate control to initiate every action. That would be exhausting. More than half of our daily actions are estimated to be automatic, accomplished without conscious deliberation. For instance, about 45 percent of our behaviors are done in the same place at the same time every day.[5] Less than half of the variation in our behavior is explained by our stated intentions.[6]

Because we often have meals or move around at the same time each day, eating and exercise are fertile ground for habit development.[7] Physical activity is under especially strong control by habits, which tell us whether to take the car or the bike to the store, whether to take the elevator or the stairs to the office, and whether it would feel weird to skip the gym today. Indeed, a lot of the overeating that's blamed on the reward system ("Obesity is increasing because most people can't resist all this tasty food") may be more correctly attributed to eating habits that became established because they produced rewards in the past but now proceed automatically. A major power of the reward system is its ability to drive habit formation.

In addition to being automatic, habits are performed without much emotion. People don't get a lot of joy from their habits, but the effort required to do them also tends to be low. For that reason alone, most of us probably don't want to make a habit of having sex or eating ice cream. On the other hand, it might be a good idea to automate some of the "should's" in our lives, the things that we know would make us feel better tomorrow or next month but don't feel satisfying in the moment, like going to bed early. Making such behaviors as frictionless as possible might be the difference between doing them and feeling guilty about not doing them. Working out mindfully could be a goal for the future, once we've figured out how to do it at all. Though we probably can't act mindfully all the time, we can choose to automate our behavior in a mindful way instead of letting habits develop by chance.

Habits of Health

The arguments presented in this book might encourage the formation of a variety of habits. A habit of daily mindfulness practice will reduce stress and improve the ability to pay attention to the body's signals of hunger and fullness—or indeed to notice anything as it's happening. A habit of regular physical activity will improve health for almost anyone. A habit of including vegetables with dinner or a piece of fruit with lunch will improve nutrition. A habit of cooking at home will make it easier to eat healthy foods. A habit of consulting with hunger before and during meals will help to avoid eating more than the body wants. Eventually a motivated person might want to make the most of these changes, but it's much easier to choose only one habit to change, as most people don't have enough willpower or self-monitoring ability to work on more than one at a time.

For a behavior to become a habit, it needs to be done in the same context every day for a while and lead to a reward. Many behaviors are self-rewarding, but feelings about particular behaviors vary across individuals. Some people feel great when they return from jogging, while others may need to end the activity with a hot shower or a chat with a friend to get a reward in the early stages.

In contrast to the conventional wisdom, forming a new habit does not take 21 days in most cases. Instead, research suggests that's about the fastest that a habit can ever be established. The number of repetitions required varies widely across individuals and activities. In one study, ninety-six participants were asked to choose one eating, drinking, or exercise behavior that they wanted to make into a habit and then report every day whether they did the activity in its chosen context (say, eating a piece of fruit with lunch) and how automatic it felt.[8] Fully automating a habit took anywhere from 18 to 254 days, with an average time of 66 days. The process was faster in people who did their chosen behavior nearly every day, but the time it took to form a habit mostly seemed to depend on how difficult the behavior was for that individual. Missing a day does not affect habit formation, but missing a week has a major effect. Exercise habits took longer to form than eating or drinking habits.

In that study and others, fewer than half of the people who decided to develop a new health habit succeeded. To improve the odds, start by taking some time to think about whether the process of making the change seems like a reasonable amount of effort. People who are uncertain if they'll succeed in forming a new habit are usually right that the plan won't work for them, so refine the plan until it inspires confidence. Explicitly considering the motivation for change also makes it more likely to happen, for example, by listing the benefits of the new habit ("Going to the gym will give me more energy") alongside the benefits of not making the change ("I'll get to sleep in longer in the morning if I don't work out").

Careful planning can help in forming the new habit.[9] The context that triggers this new behavior contributes to success or failure. As a cue, choose something that happens every day, like waking up or eating lunch. It's better to choose an event than a time as the cue because we might not notice when three in the afternoon rolls around. To bridge the gap between the cue and the behavior in the early stages of habit formation, try using a visual reminder, like leaving walking shoes on the car's passenger seat or posting a note on the mirror for a habit triggered by toothbrushing. Getting started is the important part.

Consistency is key in habit formation. Choose a behavior that's easy enough to do regularly. First form the habit of taking a ten-minute walk after dinner, then lengthen that walk or switch to jogging. Do whatever is necessary to make the new habit as convenient as possible. Keep workout clothes and equipment handy to avoid having to rummage around for them. Turn off the phone or put the cat out to prevent interruptions during meditation. Before a habit is established, any bit of annoyance can throw it off course.

Tracking whether or not the new behavior was done each day improves the odds of doing it, in part by increasing the importance of the new habit in memory. Better still, keep the tracking sheet somewhere prominent, as a reminder of the intention to do this new behavior every day. Visualizing the cue followed by the new behavior may also make it easier to remember to perform the habit. Fantasizing about how wonderful life will be after the new habit has been established, in contrast, is likely to impede progress, in part by creating a feeling that the goal has already been achieved. Mentally rehearsing the process of change rather than the outcome is the best approach.

A hard-core technique for habit change is precommitment. Like Ulysses lashing himself to the mast, precommitment is taking action when willpower is high to make it difficult to give in later when it gets weak. My Internet-blocking software is a form of precommitment. Sharing a goal and progress with others is another form of precommitment that makes it socially costly to quit. Web sites like stickK.com let people make a public commitment to a goal and add financial or reputational incentives to maintain the new behavior. For example, people can arrange for the site to donate some of their money to a charity they hate every time they fail to complete their new habit. (Like most self-help tools, the site is highly focused on weight loss, but it can be used for any goal.)

Choosing a cue that's already linked to a strong habit makes it more difficult to form the new habit. It's possible to substitute a new habit for an old one, but it's harder and requires more willpower and self-monitoring. People who want to replace an old habit should consider whether they could separate the two parts by avoiding the cue that triggers the bad habit and using a different cue for the new habit until it's fully automatic. For instance, it might be easier to drive to work by a different route to avoid passing the doughnut shop than to count on willpower to keep you from turning in. That approach will work only if it's possible to avoid the bad-habit cue forever, though.

Ditching Bad Habits

Creating a new habit is easier in many ways than breaking an old one. People are most likely to change their habits when their daily life changes, for example, during major transitions like moving to a new place, getting married, or having a baby. During these transitions, old cues disappear and new cues arise, freeing behavior to become more flexible and goal sensitive. Look for these moments and take advantage of opportunities to get rid of old habits that don't work anymore. Those moments also illustrate why we rely on habits, because the need to think about every decision all day long is exhausting. Most of us soon settle into a new set of routines, so the opportunity usually doesn't last long.

To change a habit, we need to create a new path and then use it often enough that it becomes automatic. By helping us become aware of what we do as we're doing it, mindfulness can open space for us to make different choices. Even then, the old path will still exist, and we may return to it under stress or distraction. Each successful habit change teaches us more about effective ways to approach this problem in our own lives, so habits become progressively easier to form with practice.

The most reliable way to get rid of a bad habit is to replace it with a better one by pairing the same cue with a different behavior. That's because habits always involve doing something, so it's impossible to form a habit of not doing something. All we can do is form a habit of doing something else instead. To make the change stick, the new habit must be repeated until it's as strong as the old habit, which may be a lot of times if the old habit is well practiced.

One key to changing a habit is to identify the trigger that initiates the habitual behavior. Because habits are largely unconscious, figuring out what triggers them may require some thought and experimentation. Start by keeping notes on what happened just before the habitual behavior. The cue may be a place, a time of day, an action (opening the desk drawer where the candy is stored), a sensory experience (smelling coffee or seeing the concession stand on your way into a movie), or another person. In some cases, the cue for a habit may be a thought. Once there are two or three possibilities, pay attention to which one has the most consistent relationship to the habitual behavior, or try removing one and see if the habitual behavior changes. With that information, we can work to avoid the cue or substitute another action that better suits our goals until it becomes a replacement habit.

Becoming aware of the cue as it occurs is the most challenging aspect of modifying a habit, but it's an important precondition for changing it in a way that will stick. We're often only vaguely aware of our habitual thoughts and actions, if we notice them at all, as our minds are typically elsewhere. The performance of an established habit is prime time for mind wandering because the habit commands so little attention. It's hard to notice when we're doing something we do every day and also hard to remember whether we've done it. For example, if I'm interrupted in the shower, I might stand there wondering whether I've washed my hair yet. If we don't notice the cue, though, we can't respond to it differently, and the more often we practice the old habit, the more entrenched it becomes.

Being mindful in daily life can help us to notice when the cue for a habitual behavior occurs. During the transition, it's vital to remain aware of the cue and deliberately substitute a different response to it. One approach is to rearrange the environment to place obstacles between the cue and the habit that make the bad habit more difficult to practice. Relocating the potato chips to the basement makes it hard to eat them mindlessly, and hiding the TV remote in the laundry basket gets in the way of the automatic tendency to settle on the couch at the end of the day. (Some readers may have toddlers who can perform this service for them.) Moving the alarm across the room makes it harder to hit the snooze button. A person who keeps cigarettes in the house is not yet serious about quitting smoking. Making the cue easier to recognize or the old habit harder to perform improves the odds of remembering to respond in a new way.

Habit change is likely to be more successful if we plan how we will cope with difficult situations and get back on track afterward. An effective implementation plan includes problems that may arise ("I won't want to go outside if it's raining") and what to do about them ("but I can walk around the mall instead"). A meta-analysis showed that such plans make people substantially more likely to follow through on their intentions to reach a goal by protecting them from going off track because of unexpected problems.[10] At the moment when something goes wrong, people are often tired and cranky and more interested in forgetting the whole business than in problem solving, so it helps to think about potential problems in advance. Missed opportunities to perform the new habit should lead not to recrimination but to reflection on what went wrong ("I didn't walk after work today because I went out for drinks with my team instead") and revision of the implementation plan to cover that problem ("If I have plans after work, I'll walk during lunchtime").

A common reason that people quit in the middle of a habit change is because they don't feel they're making progress. To avoid this problem, set a realistic goal and measure improvement about once a week. Try to think of a positive result of moving toward your ultimate goal that can be experienced soon, like being able to run after a child or grandchild without getting winded or avoiding that overfull feeling after a party. Goals that can be reached within weeks, like being stronger, are more motivating than goals that lie in the far future, like not getting diabetes. Instead of pursuing a long-term goal, think of health behaviors as a way of taking care of yourself to feel better now. The process of building a habit can be satisfying in its own right, as we watch the behavior become easier each day and anticipate a time when we won't have to think about it at all.

Putting the brain's energy-balance system in charge of eating follows a similar process for most people. At first remembering to respond to our internal signals requires mindful attention, but with practice the ability to sense hunger and fullness improves, until eventually it becomes hard not to notice when we're done eating. The automatic recognition of hunger frees our willpower to help us achieve new goals. Developing the ability to tune in to our bodies in a mindful way has many other positive effects. It helps us to realize when we're ill, injured, or stressed before the problem becomes too serious and find ways to care for ourselves.

Building a Better World

We all deserve more than a life of constant struggle between our intentions and our physiology. Good habits influence our behavior much more effectively than using willpower to resist temptation day after day. Instead, the effective role for willpower is to automate the routine activities that make up a healthy lifestyle, one small step at a time. Over the years, those small steps can add up to big progress. In my five years on that path, I've learned to eat when I'm hungry and stop when I'm full, as well as to exercise and meditate every day. Because those habits happen without thought, I still have plenty of capacity to give my full attention to the work and relationships that I care about. That includes pursuing my goal to use scientific evidence to promote health rather than shame, so we can create a world that welcomes people of all body types to the table.

Dennis Asbury also decided to apply his willpower to different goals. Although he still walks miles every day, he no longer counts calories. At his last checkup, his doctor gave him a clean bill of health because his blood pressure, cholesterol, and blood glucose are well within the healthy range, even though his weight has continued to drift up slowly. Since he quit dieting, he hasn't felt any urge to binge. Instead of white-knuckling his way past the kitchen, he's now putting his energy into enjoying retirement and keeping up with his children and grandchildren. He wishes that society would appreciate them for their accomplishments, instead of judging their bodies. "I have three daughters and four granddaughters," he says. "I hope the world is better for them than it has been for the young women here before."

To make that happen, we need to do more than hope. As a society, we need to build healthy attitudes around food and weight. We need to stop telling young people that their bodies aren't good enough. Indeed, we need to stop fat talk entirely, because that's a major way that the culture of weight obsession is reinforced. Most of all, we need to start giving hunger some respect, by learning to work with it rather than fighting it. No matter what we weigh, every one of us deserves to enjoy our food. If we eat mindfully, our bodies will find their way to the right weight for us, one that we can sustain by treating food as a partner in a healthy life.

Our grandparents understood how to live this way, and some people still do. As I finished this book, my husband and I paid a visit to another couple with two teenage daughters. This time the atmosphere was different. Not everyone was thin, but we were all healthy. We slept well, then went hiking or swimming for fun. No one noticed how much the girls ate or worried about whether they might gain weight. Instead our meals sparkled with laughter, talk, wine, and marvelous food. When we ate a big dinner, we skipped breakfast the next morning and then had a salad for lunch, trusting our bodies to decide how much food they wanted. When those girls grow up, I predict they won't organize their lives around the need to fit into their skinny clothes. I look forward to seeing what they'll be able to accomplish instead.

Acknowledgments

Many people were generous with their time while I was working on this book. For interviews and discussions, I owe thanks to Dennis Asbury, Linda Bacon, Sylvia Berman, Steven Blair, Ken Britten, Kelly Coffey, Elissa Epel, Yoni Freedhoff, Anne Golob, Jean Kristeller, Rudy Leibel, Andrea Lieberstein, Traci Mann, Gordon Rosenberg, Mike Schwartz, Thel Seraphim, Janet Tomiyama, Michael Vassar, and Sam Wang. For comments on the text, much appreciation to Ashley Mason, Becca Moss, Nicole Guerin, Andrea Quintero, Adele Seelke, and Sam Wang. They all helped me develop my thoughts and figure out how to present this counterintuitive information in ways that people can understand. In the end, though, this is my book, so no one else is responsible for what I chose to put into it. This area of science is contentious, and I'm sure each of them disagrees with something that I wrote.

Others helped me bring these ideas to a wide audience. At TED, thanks are due to Chris Anderson, Sarah Caddick, Bruno Giussani, Emily McManus, and the video editor with the thirteen-year-old daughter who called to me from the top of the escalator to tell me he was moved by my talk, though I didn't get his name. My agents, Lindsay Edgecombe and Jim Levine, supported this project from the initial idea to the finished book. Their ability to get behind a challenging concept and shape it into a successful proposal has been inspiring. When I can't get a book idea out of my head, it's great to know that Lindsay is equally obsessed. My editor, Niki Papadopoulos, shared my vision of what this book could be. Her guidance improved the text immensely, and her unflagging enthusiasm made the process as easy as possible. Adrian Zackheim, the publisher, was excited about these ideas from the beginning and has been an unwavering source of support.

My husband, Ken Britten, always takes good care of me—whether that means cooking dinner or challenging my arguments to make them stronger. My parents encouraged me to figure things out for myself and gave me the tools to do it well. Finally, I'm grateful to the many people I've cornered at parties to talk about these ideas over the years. Thank you all for your help.

Notes

Chapter 1: The Diet Roller Coaster

1. Gina Kolata, "How Fen-Phen, a Diet 'Miracle,' Rose and Fell," *New York Times,* September 23, 1997.
2. Alison Fildes, Judith Charlton, Caroline Rudisill, Peter Littlejohns, A. Toby Prevost, and Martin C. Gulliford, "Probability of an Obese Person Attaining Normal Body Weight: Cohort Study Using Electronic Health Records," *American Journal of Public Health* 105(9) (2015): e54–e59.
3. M. T. McGuire, R. R. Wing, and J. O. Hill, "The Prevalence of Weight Loss Maintenance Among American Adults," *International Journal of Obesity* 23(12) (1999): 1314–19.
4. Matthias B. Schulze, Teresa T. Fung, JoAnn E. Manson, Walter C. Willett, and Frank B. Hu, "Dietary Patterns and Changes in Body Weight in Women," *Obesity* 14(8) (2006): 1444–53.
5. Paula A. Quatromoni, Michael Pencina, Mark R. Cobain, Paul F. Jacques, and Ralph B. D'Agostino, "Dietary Quality Predicts Adult Weight Gain: Findings from the Framingham Offspring Study," *Obesity* 14(8) (2006): 1383–91.
6. Roland Buettner, Jürgen Schölmerich, and L. Cornelius Bollheimer, "High-Fat Diets: Modeling the Metabolic Disorders of Human Obesity in Rodents," *Obesity* 15(4) (2007): 798–808.
7. M. Rosenbaum and R. L. Leibel, "Adaptive Thermogenesis in Humans," supplement, *International Journal of Obesity* 34(1) (2010): S47–S55.
8. Aaron J. Grossberg, Jarrad M. Scarlett, and Daniel L. Marks, "Hypothalamic Mechanisms in Cachexia," *Physiology & Behavior* 100(5) (2010): 478–89.

9. C. Filozof, M. C. Fernández Pinilla, and A. Fernández-Cruz, "Smoking Cessation and Weight Gain," *Obesity Reviews* 5(2) (2004): 95–103.

10. Katherine M. Flegal, Richard P. Troiano, Elsie R. Pamuk, Robert K. Kuczmarski, and Stephen M. Campbell, "The Influence of Smoking Cessation on the Prevalence of Overweight in the United States," *New England Journal of Medicine* 333(18) (1995): 1165–70.

11. Margaret A. Stefater, Diego Pérez-Tilve, Adam P. Chambers, Hilary E. Wilson-Pérez, Darleen A. Sandoval, José Berger, Mouhamadoul Toure, et al., "Sleeve Gastrectomy Induces Loss of Weight and Fat Mass in Obese Rats, but Does Not Affect Leptin Sensitivity," *Gastroenterology* 138(7) (2010): 2426–36.

12. J. A. Tadross and C. W. le Roux, "The Mechanisms of Weight Loss After Bariatric Surgery," supplement, *International Journal of Obesity* 33(1) (2009): S28–S32.

13. Lars Sjöström, Kristina Narbro, C. David Sjöström, Kristjan Karason, Bo Larsson, Hans Wedel, Ted Lystig, et al., "Effects of Bariatric Surgery on Mortality in Swedish Obese Subjects," *New England Journal of Medicine* 357(8) (2007): 741–52.

14. Henry Buchwald, Rhonda Estok, Kyle Fahrbach, Deirdre Banel, and Isabella Sledge, "Trends in Mortality in Bariatric Surgery: A Systematic Review and Meta-analysis," *Surgery* 142(4) (2007): 621–35.

15. Heena P. Santry, Daniel L. Gillen, and Diane S. Lauderdale, "Trends in Bariatric Surgical Procedures," *Journal of the American Medical Association* 294(15) (2005): 1909–17.

16. David S. Zingmond, Marcia L. McGory, and Clifford Y. Ko, "Hospitalization Before and After Gastric Bypass Surgery," *Journal of the American Medical Association* 294(15) (2005): 1918–24.

17. L. Sjöström, "Review of the Key Results from the Swedish Obese Subjects (SOS) Trial—A Prospective Controlled Intervention Study of Bariatric Surgery," *Journal of Internal Medicine* 273 (2013): 219–34.

18. Stuart Pfeifer, "A Second Lap-Band Firm Is Warned," *Los Angeles Times,* July 31, 2012.

19. Santry, Gillen, and Lauderdale, "Trends in Bariatric Surgical Procedures."

20. Yoon K. Loke, Sheena Derry, and Angharad Pritchard-Copley, "Appetite

Suppressants and Valvular Heart Disease—A Systematic Review," *BMC Clinical Pharmacology* 2 (2002): 6.

21. Stephanie Saul, "Fen-Phen Case Lawyers Say They'll Reject Wyeth Offer," *New York Times,* February 17, 2005.

22. Coulter Jones, John Fauber, and Kristina Fiore, "Diet Drug Companies Spend $60 Million Since 2010 amid Push to Win FDA Approval for New Products," *Milwaukee Journal Sentinel,* April 19, 2015.

23. Sayeed Ikramuddin, Robin P. Blackstone, Anthony Brancatisano, James Toouli, Sajani N. Shah, Bruce M. Wolfe, Ken Fujioka, et al., "Effect of Reversible Intermittent Intra-abdominal Vagal Nerve Blockade on Morbid Obesity: The ReCharge Randomized Clinical Trial," *Journal of the American Medical Association* 312(9) (2014): 915–22.

24. Maureen T. McGuire, Rena R. Wing, Mary L. Klem, Wei Lang, and James O. Hill, "What Predicts Weight Regain in a Group of Successful Weight Losers?" *Journal of Consulting and Clinical Psychology* 67(2) (1999): 177–85.

25. Michael L. Dansinger, Athina Tatsioni, John B. Wong, Mei Chung, and Ethan M. Balk, "Meta-analysis: The Effect of Dietary Counseling for Weight Loss," *Annals of Internal Medicine* 147(1) (2007): 41–50.

26. C. Ayyad and T. Andersen, "Long-Term Efficacy of Dietary Treatment of Obesity: A Systematic Review of Studies Published Between 1931 and 1999," *Obesity Reviews* 1(2) (2000): 113–19.

27. MarketData Enterprises, quoted in Andrew J. Hill, "Does Dieting Make You Fat?" supplement, *British Journal of Nutrition* 92(1) (2004): S15–S8.

28. Traci Mann, A. Janet Tomiyama, Erika Westling, Ann-Marie Lew, Barbra Samuels, and Jason Chatman, "Medicare's Search for Effective Obesity Treatments: Diets Are Not the Answer," *American Psychologist* 62(3) (2007): 220–33.

29. Melbourne F. Hovell, Alma Koch, C. Richard Hofstetter, Carol Sipan, Patricia Faucher, Ann Dellinger, Gerald Borok, et al., "Long-Term Weight Loss Maintenance: Assessment of a Behavioral and Supplemented Fasting Regimen," *American Journal of Public Health* 78(6) (1988): 663–68.

30. Francine Grodstein, Rachel Levine, Lisa Troy, Terri Spencer, Graham A. Colditz, and Meir J. Stampfer, "Three-Year Follow-up of Participants in a

Commercial Weight Loss Program: Can You Keep It Off?" *Archives of Internal Medicine* 156(12) (1996): 1302–5.

31. F. M. Kramer, R. W. Jeffery, J. L. Forster, and M. K. Snell, "Long-Term Follow-up of Behavioral Treatment for Obesity: Patterns of Weight Regain Among Men and Women," *International Journal of Obesity* 13(2) (1989): 123–36.

32. Frank M. Sacks, George A. Bray, Vincent J. Carey, Steven R. Smith, Donna H. Ryan, Stephen D. Anton, Katherine McManus, et al., "Comparison of Weight-Loss Diets with Different Compositions of Fat, Protein, and Carbohydrates," *New England Journal of Medicine* 360(9) (2009): 859–73.

33. Lisa M. Nackers, Kathryn M. Ross, and Michael G. Perri, "The Association Between Rate of Initial Weight Loss and Long-Term Success in Obesity Treatment: Does Slow and Steady Win the Race?" *International Journal of Behavioral Medicine* 17(3) (2010): 161–67.

34. Stephen Whybrow, Claus Mayer, Terry R. Kirk, Nik Mazlan, and R. James Stubbs, "Effects of Two Weeks' Mandatory Snack Consumption on Energy Intake and Energy Balance," *Obesity* 15(3) (2007): 673–85.

35. Diabetes Prevention Program Research Group, "10-Year Follow-up of Diabetes Incidence and Weight Loss in the Diabetes Prevention Program Outcomes Study," *Lancet* 374(9702) (2009): 1677–86.

36. The Look AHEAD Research Group, "Eight-Year Weight Losses with an Intensive Lifestyle Intervention: The Look AHEAD Study," *Obesity* 22(1) (2014): 5–13.

37. Ibid.

38. J.-P. Montani, Y. Schutz, and A. G. Dulloo, "Dieting and Weight Cycling as Risk Factors for Cardiometabolic Diseases: Who Is Really at Risk?" supplement, *Obesity Reviews* 16(S1) (2015): 71–78.

39. S. Yaemsiri, M. M. Slining, and S. K. Agarwal, "Perceived Weight Status, Overweight Diagnosis, and Weight Control Among US Adults: The NHANES 2003–2008 Study," *International Journal of Obesity* 35(8) (2011): 1063–70.

40. A. G. Dulloo, J. Jacquet, J.-P. Montani, and Y. Schutz, "How Dieting Makes the Lean Fatter: From a Perspective of Body Composition Autoregulation

Through Adipostats and Proteinstats Awaiting Discovery," supplement, *Obesity Reviews* 16(S1) (2015): 25–35.

41. Steven N. Blair, "Physical Inactivity: The Biggest Public Health Problem of the 21st Century," *British Journal of Sports Medicine* 43(1) (2009): 1–2.

42. Paula M. Lantz, Ezra Golberstein, James S. House, and Jeffrey Morenoff, "Socioeconomic and Behavioral Risk Factors for Mortality in a National 19-Year Prospective Study of U.S. Adults," *Social Science & Medicine* 70(10) (2010): 1558–66.

43. Julianne Holt-Lunstad, Timothy B. Smith, and J. Bradley Layton, "Social Relationships and Mortality Risk: A Meta-analytic Review," *PLoS Medicine* 7(7) (2010): e1000316.

44. Anjali Jain, "Treating Obesity in Individuals and Populations," *British Medical Journal* 331 (2005): 1387–90.

Chapter 2: Willpower Runs Out

1. Jonathan D. Hommel, Richard Trinko, Robert M. Sears, Dan Georgescu, Zong-Wu Liu, Xiao-Bing Gao, Jeremy J. Thurmon, et al., "Leptin Receptor Signaling in Midbrain Dopamine Neurons Regulates Feeding," *Neuron* 51(6) (2006): 801–10.

2. Michael Rosenbaum, Melissa Sy, Katherine Pavlovich, Rudolph L. Leibel, and Jules Hirsch, "Leptin Reverses Weight Loss–Induced Changes in Regional Neural Activity Responses to Visual Food Stimuli," *Journal of Clinical Investigation* 118(7) (2008): 2583–91.

3. I. Sadaf Farooqi, Edward Bullmore, Julia Keogh, Jonathan Gillard, Stephen O'Rahilly, and Paul C. Fletcher, "Leptin Regulates Striatal Regions and Human Eating Behavior," *Science* 317(5843) (2007): 1355.

4. Stephanie Fulton, Barbara Woodside, and Peter Shizgal, "Modulation of Brain Reward Circuitry by Leptin," *Science* 287(5450) (2000): 125–28.

5. Joy Hao, Soledad Cabeza de Vaca, Yan Pan, and Kenneth D. Carr, "Effects of Central Leptin Infusion on the Reward-Potentiating Effect of D-Amphetamine," *Brain Research* 1087(1) (2006): 123–33.

6. Ann E. Kelley and Kent C. Berridge, "The Neuroscience of Natural Re-

wards: Relevance to Addictive Drugs," *Journal of Neuroscience* 22(9) (2002): 3306–11.

7. Magalie Lenoir, Fuschia Serre, Lauriane Cantin, and Serge H. Ahmed, "Intense Sweetness Surpasses Cocaine Reward," *PLoS ONE* 2(8) (2007): e698.
8. Brian Wansink and Jeffrey Sobal, "Mindless Eating: The 200 Daily Food Decisions We Overlook," *Environment and Behavior* 39(1) (2007): 106–23.
9. Adela R. Moldovan and Daniel David, "Features of Automaticity in Eating Behavior," *Eating Behaviors* 13(1) (2012): 46–48.
10. Edward H. Nieh, Gillian A. Matthews, Stephen A. Allsop, Kara N. Presbrey, Christopher A. Leppla, Romy Wichmann, Rachael Neve, et al., "Decoding Neural Circuits That Control Compulsive Sucrose Seeking," *Cell* 160(3) (2015): 528–41.
11. Adele Diamond, "Executive Functions," *Annual Review of Psychology* 64 (2013): 135–68.
12. A. Janet Tomiyama, Traci Mann, Danielle Vinas, Jeffrey Hunger, Jill DeJager, and Shelley E. Taylor, "Low-Calorie Dieting Increases Cortisol," *Psychosomatic Medicine* 72(4) (2010): 357–64.
13. Roy F. Baumeister, Ellen Barslavsky, Mark Muraven, and Dianne M. Tice, "Ego Depletion: Is the Active Self a Limited Resource?" *Journal of Personality and Social Psychology* 74(5) (1998): 1252–65.
14. Denise T. D. de Ridder, Gerty Lensvelt-Mulders, Catrin Finkenauer, F. Marijn Stok, and Roy F. Baumeister, "Taking Stock of Self-Control: A Meta-analysis of How Trait Self-Control Relates to a Wide Range of Behaviors," *Personality and Social Psychology Review* 16(1) (2012): 76–99.
15. David T. Neal, Wendy Wood, and Aimee Drolet, "How Do People Adhere to Goals When Willpower Is Low? The Profits (and Pitfalls) of Strong Habits," *Journal of Personality and Social Psychology* 104(6) (2013): 959–75.
16. Catherine L. Davis, Phillip D. Tomporowski, Jennifer E. McDowell, Benjamin P. Austin, Patricia H. Miller, Nathan E. Yanasak, Jerry D. Allison, and Jack A. Naglierei, "Exercise Improves Executive Function and Achievement and Alters Brain Activation in Overweight Children: A Randomized, Controlled Trial," *Health Psychology* 30(1) (2011): 91–98.

Chapter 3: How Diets Lead to Weight Gain

1. Todd Tucker, *The Great Starvation Experiment: Ancel Keys and the Men Who Starved for Science* (Minneapolis: University of Minnesota Press, 2006).

2. Janet Polivy, Sharon B. Zeitlin, C. Peter Herman, and A. Lynne Beal, "Food Restriction and Binge Eating: A Study of Former Prisoners of War," *Journal of Abnormal Psychology* 103(2) (1994): 409–11.

3. Jules Hirsch, "Obesity: Matter Over Mind?" *Cerebrum*, January 1, 2003.

4. A. G. Dulloo, J. Jacquet, J.-P. Montani, and Y. Schutz, "How Dieting Makes the Lean Fatter: From a Perspective of Body Composition Autoregulation Through Adipostats and Proteinstats Awaiting Discovery," supplement, *Obesity Reviews* 16(S1) (2015): 25–35.

5. R. S. MacLean, J. A. Higgins, E. D. Giles, V. D. Sherk, and M. R. Jackman, "The Role for Adipose Tissue in Weight Regain After Weight Loss," supplement, *Obesity Reviews* 16(S1) (2015): 45–54.

6. M. R. Lowe, "Dieting: Proxy or Cause of Future Weight Gain?" supplement, *Obesity Reviews* 16(S1) (2015): 19–24.

7. K. H. Pietiläinen, S. E. Saarni, J. Kaprio, and A. Rissanen, "Does Dieting Make You Fat? A Twin Study," *International Journal of Obesity* 36(3) (2012): 456–64.

8. S. E. Saarni, A. Rissanen, S. Sarna, M. Koskenvuo, and J. Kaprio, "Weight Cycling of Athletes and Subsequent Weight Gain in Middleage," *International Journal of Obesity* 30(11) (2006): 1639–44.

9. A. Janet Tomiyama, Traci Mann, Danielle Vinas, Jeffrey Hunger, Jill DeJager, and Shelley E. Taylor, "Low-Calorie Dieting Increases Cortisol," *Psychosomatic Medicine* 72(4) (2010): 357–64.

10. Eva Kemps, Marika Tiggemann, and Kelly Marshall, "Relationship Between Dieting to Lose Weight and the Functioning of the Central Executive," *Appetite* 45(3) (2005): 287–94.

11. Eva Kemps and Marika Tiggemann, "Working Memory Performance and Preoccupying Thoughts in Female Dieters: Evidence for a Selective Central Executive Impairment," *British Journal of Clinical Psychology* 44(3) (2005): 357–66.

12. Ingrid C. Fedoroff, Janet Polivy, and C. Peter Herman, "The Effect of

Pre-exposure to Food Cues on the Eating Behavior of Restrained and Unrestrained Eaters," *Appetite* 28(1) (1997): 33–47.

13. Eric Stice, Katherine Presnell, Heather Shaw, and Paul Rohde, "Psychological and Behavioral Risk Factors for Obesity Onset in Adolescent Girls: A Prospective Study," *Journal of Consulting and Clinical Psychology* 73(2) (2005): 195–202.
14. S. A. French, R. W. Jeffery, J. L. Forster, P. G. McGovern, S. H. Keider, and J. E. Baxter, "Predictors of Weight Change Over Two Years Among a Population of Working Adults: The Healthy Worker Project," *International Journal of Obesity and Related Metabolic Disorders* 18(3) (1994): 145–54.
15. Reinhard J. Tuschl, "From Dietary Restraint to Binge Eating: Some Theoretical Considerations," *Appetite* 14(2) (1990): 105–9.
16. Janet Polivy, "Psychological Consequences of Food Restriction," *Journal of the American Dietetic Association* 96(6) (2006): 589–92.
17. Ellen A. Schur, Susan R. Heckbert, and Jack H. Goldberg, "The Association of Restrained Eating with Weight Change Over Time in a Community-Based Sample of Twins," *Obesity* 18(6) (2010): 1146–52.
18. R. R. Fabsitz, P. Sholinsky, and D. Carmelli, "Genetic Influences on Adult Weight Gain and Maximum Body Mass Index in Male Twins," *American Journal of Epidemiology* 140(8) (1994): 711–20.
19. Sarah E. Racine, S. Alexandra Burt, William G. Iacono, Matt McGue, and Kelly L. Klump, "Dietary Restraint Moderates Genetic Risk for Binge Eating," *Journal of Abnormal Psychology* 120(1) (2011): 119–28.
20. Eric Stice, Kendra Davis, Nicole P. Miller, and C. Nathan Marti, "Fasting Increases Risk for Onset of Binge Eating and Bulimic Pathology: A 5-Year Prospective Study," *Journal of Abnormal Psychology* 117(4) (2008): 941–46.
21. Rebecca L. Corwin, Nicole M. Avena, and Mary M. Boggiano, "Feeding and Reward: Perspectives from Three Rat Models of Binge Eating," *Physiology & Behavior* 104(1) (2011): 87–97.
22. Mary M. Hagan, Paula C. Chandler, Pamela K. Wauford, Rachel J. Rybak, and Kimberly D. Oswald, "The Role of Palatable Food and Hunger as Trigger Factors in an Animal Model of Stress Induced Binge Eating," *International Journal of Eating Disorders* 34(2) (2003): 183–97.

23. Kenneth D. Carr, "Food Scarcity, Neuroadaptations, and the Pathogenic Potential of Dieting in an Unnatural Ecology: Binge Eating and Drug Abuse," *Physiology & Behavior* 104(1) (2011): 16–27.

24. Hisham Ziauddeen, I. Sadaf Farooqi, and Paul C. Fletcher, "Obesity and the Brain: How Convincing Is the Addiction Model?" *Nature Reviews Neuroscience* 13(4) (2012): 279–86.

25. Michael W. Wiederman and Tamara Pryor, "Substance Abuse and Impulsive Behaviors Among Adolescents with Eating Disorders," *Addictive Behavior* 21(2) (1996): 269–72.

26. Lisa R. R. Lilenfeld, Rebecca Ringham, Melissa A. Kalarchian, and Marsha D. Marcus, "A Family History Study of Binge-Eating Disorder," *Comprehensive Psychiatry* 49(3) (2008): 247–54.

Chapter 4: The Weight of Beauty

1. Anne E. Becker, Rebecca A. Burwell, Stephen E. Gilman, David B. Herzog, and Paul Hamburg, "Eating Behaviours and Attitudes Following Prolonged Exposure to Television Among Ethnic Fijian Adolescent Girls," *British Journal of Psychiatry* 180(6) (2002): 509–14.

2. Janet Polivy and C. Peter Herman, "Causes of Eating Disorders," *Annual Review of Psychology* 53 (2002):187–213.

3. Shelly Grabe, L. Monique Ward, and Janet Shibley Hyde, "The Role of the Media in Body Image Concerns Among Women: A Meta-analysis of Experimental and Correlational Studies," *Psychological Bulletin* 134(3) (2008): 460–76.

4. Harrison G. Pope, Roberto Olivardia, Amanda Gruber, and John Borowiecki, "Evolving Ideals of Male Body Image as Seen Through Action Toys," *International Journal of Eating Disorders* 26(1) (1999): 65–72.

5. Alan Feingold and Ronald Mazzella, "Gender Differences in Body Image Are Increasing," *Psychological Science* 9(3) (1998): 190–95.

6. Sarah Grogan, *Body Image: Understanding Body Dissatisfaction in Men, Women, and Children* (New York: Routledge, 2008).

7. Grabe, Ward, and Hyde, "The Role of the Media in Body Image Concerns Among Women."

8. Jeffrey M. Hunger and A. Janet Tomiyama, "Weight Labeling and Obesity: A Longitudinal Study of Girls Aged 10 to 19 Years," *JAMA Pediatrics* 168(6) (2014): 579–80.

9. Daniel Callahan, "Obesity: Chasing an Elusive Epidemic," *Hastings Center Report* 43(1) (2013): 34–40.

10. Phebe Cramer and Tiffany Steinwert, "Thin Is Good, Fat Is Bad: How Early Does It Begin?" *Journal of Applied Developmental Psychology* 19(3) (1998): 429–51.

11. R. M. Puhl, T. Andreyeva, and K. D. Brownell, "Perceptions of Weight Discrimination: Prevalence and Comparison to Race and Gender Discrimination in America," *International Journal of Obesity* 32(6) (2008): 992–1000.

12. Rebecca M. Puhl and Kelly D. Brownell, "Confronting and Coping with Weight Stigma: An Investigation of Overweight and Obese Adults," *Obesity* 14(10) (2006): 1802–15.

13. Cheri L. Olson, Howard D. Schumake, and Barbara P. Yawn, "Overweight Women Delay Medical Care," *Archives of Family Medicine* 3(10) (1994): 888–92.

14. Angelina R. Sutin, Yannick Stephan, and Antonio Terracciano, "Weight Discrimination and Risk of Mortality," *Psychological Science* (2015), online ahead of print. DOI: 10.1177/0956797615601103.

15. Angelina R. Sutin and Antonio Terracciano, "Perceived Weight Discrimination and Obesity," *PLoS ONE* 8(7) (2013): e70048.

16. Sarah E. Jackson, Rebecca J. Beeken, and Jane Wardle, "Perceived Weight Discrimination and Changes in Weight, Waist Circumference, and Weight Status," *Obesity* 22(12) (2014): 2485–88.

17. Lenny R. Vartanian and Sarah A. Novak, "Internalized Societal Attitudes Moderate the Impact of Weight Stigma on Avoidance of Exercise," *Obesity* 19(4) (2011): 757–62.

18. Brenda Major, Jeffrey M. Hunger, Debra P. Bunyan, and Carol T. Miller, "The Ironic Effects of Weight Stigma," *Journal of Experimental Social Psychology* 51 (2014): 74–80.

19. Mark L. Hatzenbuehler, Katherine M. Keyes, and Deborah S. Hasin, "Associations Between Perceived Weight Discrimination and the Prevalence of Psychiatric Disorders in the General Population," *Obesity* 17(11) (2009): 2033–39.

20. Vartanian and Novak, "Internalized Societal Attitudes Moderate the Impact of Weight Stigma on Avoidance of Exercise."

21. Paul T. Williams, "Self-Selection Accounts for Inverse Association Between Weight and Cardiorespiratory Fitness," *Obesity* 16(1) (2008): 102–6.

22. Dianne Neumark-Sztainer, Melanie Wall, Jia Guo, Mary Story, Jess Haines, and Maria Eisenberg, "Obesity, Disordered Eating, and Eating Disorders in a Longitudinal Study of Adolescents: How Do Dieters Fare Five Years Later?" *Journal of the American Dietetic Association* 106(4) (2006): 559–68.

23. Dianne Neumark-Sztainer, Melanie Wall, Mary Story, and Amber R. Standish, "Dieting and Unhealthy Weight Control Behaviors During Adolescence: Associations with 10-Year Changes in Body Mass Index," *Journal of Adolescent Health* 50(1) (2012): 80–86.

24. Alison E. Field, S. B. Austin, C. B. Taylor, Susan Malspeis, Bernard Rosner, Helaine R. Rockett, Matthew W. Gillman and Graham A. Colditz, "Relation Between Dieting and Weight Change Among Preadolescents and Adolescents," *Pediatrics* 112(4) (2003): 900–906.

25. Koenraad Cuypers, Kirsti Kvaløy, Grete Bratberg, Kristian Midthjell, Jostein Holmen, and Turid Lingaas Holman, "Being Normal Weight but Feeling Overweight in Adolescence May Affect Weight Development into Young Adulthood—An 11-Year Followup: The HUNT Study, Norway," *Journal of Obesity* 2012 (2012): 601872.

26. Dianne Neumark-Sztainer, Melanie Wall, Jess Haines, Mary Story, and Maria E. Eisenberg, "Why Does Dieting Predict Weight Gain in Adolescents? Findings from Project EAT-II: A 5-Year Longitudinal Study," *Journal of the American Dietetic Association* 107(3) (2007): 448–55.

27. Eric Stice, Rebecca P. Cameron, Joel D. Killen, Chris Hayward, and C. Barr Taylor, "Naturalistic Weight-Reduction Efforts Prospectively Predict Growth in Relative Weight and Onset of Obesity Among Female Adolescents," *Journal of Consulting and Clinical Psychology* 67(6) (1999): 967–74.

28. Eric Stice, Chris Hayward, Rebecca P. Cameron, Joel D. Killen, and C. Barr Taylor, "Body-Image and Eating Disturbances Predict Onset of Depression Among Female Adolescents: A Longitudinal Study," *Journal of Abnormal Psychology* 109(3) (2000): 438–44.

29. Katherine N. Balantekin, Jennifer S. Savage, Michele E. Marini, and Leann L. Birch, "Parental Encouragement of Dieting Promotes Daughters' Early Dieting," *Appetite* 80(1) (2014): 190–96.

30. Nadia Micali, George Ploubidis, Bianca De Stavola, Emily Simonoff, and Janet Treasure, "Frequency and Patterns of Eating Disorder Symptoms in Early Adolescence," *Journal of Adolescent Health* 54(5) (2014): 574–81.

31. Sonja A. Swanson, Scott J. Crow, Daniel Le Grange, Joel Swendsen, and Kathleen R. Merikangas, "Prevalence and Correlates of Eating Disorders in Adolescents: Results from the National Comorbidity Survey Replication Adolescent Supplement," *Archives of General Psychiatry* 68(7) (2011): 714–23.

32. Catherine M. Shisslak, Marjorie Crago, and Linda S. Estes, "The Spectrum of Eating Disturbances," *International Journal of Eating Disorders* 18(3) (1995): 209–19.

33. Frédérique R. E. Smink, Daphne van Hoeken, and Hans W. Hoek, "Epidemiology, Course, and Outcome of Eating Disorders," *Current Opinion in Psychiatry* 26(6) (2013): 543–48.

34. Anu Raevuori, Anna Keski-Rahkonen, and Hans W. Hoek, "A Review of Eating Disorders in Males," *Current Opinion in Psychiatry* 27(6) (2014): 426–30.

35. Eric Stice and Heather E. Shaw, "Role of Body Dissatisfaction in the Onset and Maintenance of Eating Pathology: A Synthesis of Research Findings," *Journal of Psychosomatic Research* 53(5) (2002): 985–93.

36. The Body Project, www.bodyprojectsupport.org/home.

37. Eric Stice, C. Nathan Marti, Sonja Spoor, Katherine Presnell, and Heather Shaw, "Dissonance and Healthy Weight Eating Disorder Prevention Programs: Long-Term Effects from a Randomized Efficacy Trial," *Journal of Consulting and Clinical Psychology* 76(2) (2008): 329–40.

38. eBody Project, www.ebodyproject.org/.

39. Eric Stice, Shelley Durant, Paul Rohde, and Heather Shaw, "Effects of a Prototype Internet Dissonance-Based Eating Disorder Prevention Program at 1- and 2-Year Follow-up," *Health Psychology* 33(12) (2014): 1558–67.

Chapter 5: Our Brains Fight Weight Loss

1. Retro Report, "Biosphere 2: An American Space Odyssey," *New York Times,* June 10, 2013.

2. Christian Weyer, Roy L. Walford, Inge T. Harper, Mike Milner, Taber MacCallum, P. Antonio Tataranni, and Eric Ravussin, "Energy Metabolism After 2 Y of Energy Restriction: The Biosphere 2 Experiment," *American Journal of Clinical Nutrition* 72(4) (2000): 946–53.

3. Rudolph L. Leibel and Jules Hirsch, "Diminished Energy Requirements in Reduced-Obese Patients," *Metabolism* 33(2) (1984): 164–70.

4. Rudolph L. Leibel, Michael Rosenbaum, and Jules Hirsch, "Changes in Energy Expenditure Resulting from Altered Body Weight," *New England Journal of Medicine* 332(10) (1995): 621–28.

5. Michael Rosenbaum, Jules Hirsch, Dympna A. Gallagher, and Rudolph L. Leibel, "Long-Term Persistence of Adaptive Thermogenesis in Subjects Who Have Maintained a Reduced Body Weight," *American Journal of Clinical Nutrition* 88(4) (2008): 906–12.

6. Gina Kolata, *Rethinking Thin: The New Science of Weight Loss—and the Myths and Realities of Dieting* (New York: Picador, 2007).

7. Darcy L. Johannsen, Nicholas D. Knuth, Robert Huizenga, Jennifer C. Rood, Eric Ravussin, and Kevin D. Hall, "Metabolic Slowing with Massive Weight Loss Despite Preservation of Fat-Free Mass," *Journal of Clinical Endocrinology & Metabolism* 97(7) (2012): 2489–96.

8. D. B. West, D. Fey, and S. C. Woods, "Cholecystokinin Persistently Suppresses Meal Size but Not Food Intake in Free-Feeding Rats," *American Journal of Physiology* 246(5) (1984): R776–R787.

9. Laurence E. Ring and Lori M. Zeltser, "Disruption of Hypothalamic Leptin Signaling in Mice Leads to Early-Onset Obesity, but Physiological Adapta-

tions in Mature Animals Stabilize Adiposity Levels," *Journal of Clinical Investigation* 120(8) (2010): 2931–41.

10. Michael Rosenbaum, Rochelle Goldsmith, Daniel Bloomfield, Anthony Magnano, Louis Weimer, Steven Heymsfield, Dympna Gallagher, et al., "Low Dose Leptin Reverses Skeletal Muscle, Autonomic, and Neuroendocrine Adaptations to Maintenance of Reduced Weight," *Journal of Clinical Investigation* 115(12) (2005): 3579–86.
11. Xiaoqing Zhang, Guo Zhang, Hai Zhang, Michael Karin, Hua Bai, and Dongsheng Cai, "Hypothalamic IKKβ/NFκB and ER Stress Link Overnutrition to Energy Imbalance and Obesity," *Cell* 135(1) (2008): 61–73.
12. Joshua P. Thaler, Chun-Xia Yi, Ellen A. Schur, Stephan J. Guyenet, Bang H. Hwang, Marcelo O. Dietrich, Xiaolin Zho, et al., "Obesity Is Associated with Hypothalamic Injury in Rodents and Humans," *Journal of Clinical Investigation* 122(1) (2012): 153–62.
13. Marcelo A. Mori, Manway Liu, Olivier Bezy, Katrine Almind, Hagit Shapiro, Simon Kasif, and C. Ronald Kahn, "A Systems Biology Approach Identifies Inflammatory Abnormalities Between Mouse Strains Prior to Development of Metabolic Disease," *Diabetes* 59(11) (2010): 2960–71.
14. Aruna D. Pradhan, JoAnn E. Manson, Nader Rifai, Julie E. Buring, and Paul M. Ridker, "C-Reactive Protein, Interleukin 6, and Risk of Developing Type 2 Diabetes Mellitus," *Journal of the American Medical Association* 286(3) (2001): 327–34.
15. Joshua P. Thaler, Chun-Xia Yi, Ellen A. Schur, Stephan J. Guyenet, Bang H. Hwang, Marcelo O. Dietrich, Xiaolin Zhao, et al., "Obesity Is Associated with Hypothalamic Injury in Rodents and Humans," *Journal of Clinical Investigation* 122(1) (2012): 153–62.
16. Deniz Atasoy, J. Nicholas Betley, Helen H. Su, and Scott M. Sternson, "Deconstruction of a Neural Circuit for Hunger," *Nature* 488(7410) (2012): 172–77.
17. Michael J. Krashes, Shuichi Koda, ChianPing Ye, Sarah C. Rogan, Andrew C. Adams, Daniel S. Cusher, Eleftheria Maratos-Flier, et al., "Rapid, Reversible Activation of AgRP Neurons Drives Feeding Behavior in Mice," *Journal of Clinical Investigation* 121(4) (2011): 1424–28.

Chapter 6: Early Life and Adult Weight

1. Gian-Paolo Ravelli, Zema A. Stein, and Mervyn W. Susser, "Obesity in Young Men After Famine Exposure in Utero and Early Infancy," *New England Journal of Medicine* 295(7) (1976): 349–53.

2. L. H. Lumey, Aryeh D. Stein, and Ezra Susser, "Prenatal Famine and Adult Health," *Annual Review of Public Health* 32 (2011): 237–62.

3. Patrick Bateson, David Barker, Timothy Clutton-Brock, Debal Deb, Bruno D'Udine, Robert A. Foley, Peter Gluckman, et al., "Developmental Plasticity and Human Health," *Nature* 430(6998) (2004): 419–21.

4. W. C. Knowler, D. J. Pettitt, M. F. Saad, M. A. Charles, R. G. Nelson, B. V. Howard, C. Bogardus, and P. H. Bennett, "Obesity in the Pima Indians: Its Magnitude and Relationship with Diabetes," supplement, *American Journal of Clinical Nutrition* 53(6) (1991): 1543S–1551S.

5. Michael S. Kramer, Isabelle Morin, Hong Yang, Robert W. Platt, Robert Usher, Helen McNamara, K. S. Joseph, and Shi Wu Wen, "Why Are Babies Getting Bigger? Temporal Trends in Fetal Growth and Its Determinants," *Journal of Pediatrics* 141(4) (2002): 538–42.

6. Janis Baird, David Fisher, Patricia Lucas, Jos Kleijnen, Helen Roberts, and Catherine Law, "Being Big or Growing Fast: Systematic Review of Size and Growth in Infancy and Later Obesity," *British Medical Journal* 331(7522) (2005): 929.

7. Robert C. Whitaker, Jeffrey A. Wright, Margaret S. Pepe, Kristy D. Seidel, and William H. Dietz, "Predicting Obesity in Young Adulthood from Childhood and Parental Obesity," *New England Journal of Medicine* 337(13) (1997): 869–73.

8. David S. Ludwig and Janet Currie, "The Association Between Pregnancy Weight Gain and Birthweight: A Within-Family Comparison," *Lancet* 376(9745) (2010): 984–90.

9. Sebastien G. Bouret, "Neurodevelopmental Actions of Leptin," *Brain Research* 1350 (2010): 2–9.

10. Merly C. Vogt, Lars Paeger, Simon Hess, Sophie M. Steculorum, Motoharu Awazawa, Brigitte Hampel, Susanne Neupert, et al., "Neonatal Insulin

Action Impairs Hypothalamic Neurocircuit Formation in Response to Maternal High-Fat Feeding," *Cell* 156(3) (2014): 495–509.

11. Shona L. Kirk, Anne-Maj Samuelsson, Marco Argenton, Hannah Dhonye, Theodosis Kalamatianos, Lucilla Poston, Paul D. Taylor, and Clive W. Coen, "Maternal Obesity Induced by Diet in Rats Permanently Influences Central Processes Regulating Food Intake in Offspring," *PLoS ONE* 4(6) (2009): e5870.
12. Sophie M. Steculorum and Sebastien G. Bouret, "Maternal Diabetes Compromises the Organization of Hypothalamic Feeding Circuits and Impairs Leptin Sensitivity in Offspring," *Endocrinology* 152(11) (2011): 4171–79.
13. J. L. Kaar, J. T. Brinton, T. Crume, R. F. Hamman, D. H. Glueck, and D. Dabelea, "Leptin Levels at Birth and Infant Growth: The EPOCH Study," *Journal of Developmental Origins of Health and Disease* 5(3) (2014): 214–18.
14. Daniel A. Lee, Joseph L. Bedont, Thomas Pak, Hong Wang, Juan Song, Ana Miranda-Angulo, Vani Takiar, et al., "Tanycytes of the Hypothalamic Median Eminence Form a Diet-Responsive Neurogenic Niche," *Nature Neuroscience* 15(5) (2012): 700–702.
15. D. F. Williamson, T. J. Thompson, R. F. Anda, W. H. Dietz, and V. Felitti, "Body Weight and Obesity in Adults and Self-Reported Abuse in Childhood," *International Journal of Obesity and Related Metabolic Disorders* 26(8) (2002): 1075–82.
16. I. Lissau and T. I. A. Sørensen, "Parental Neglect During Childhood and Increased Risk of Obesity in Young Adulthood," *Lancet* 343(8893) (1994): 324–27.
17. I. Lissau and T. I. Sørensen, "School Difficulties in Childhood and Risk of Overweight and Obesity in Young Adulthood: A Ten Year Prospective Population Study," *International Journal of Obesity and Related Metabolic Disorders* 17(3) (1993): 169–75.
18. A. Danese and M. Tan, "Childhood Maltreatment and Obesity: Systematic Review and Meta-analysis," *Molecular Psychiatry* 19(5) (2014): 544–54.
19. E. Oken, E. B. Levitan, and M. W. Gillman, "Maternal Smoking During Pregnancy and Child Overweight: Systematic Review and Meta-analysis," *International Journal of Obesity* 32(2) (2008): 201–10.

20. Boris Novakovic, Joanne Ryan, Natalie Pereira, Berin Boughton, Jeffrey M. Craig, and Richard Saffery, "Postnatal Stability, Tissue, and Time Specific Effects of AHRR Methylation Change in Response to Maternal Smoking in Pregnancy," *Epigenetics* 9(3) (2014): 377–86.

21. Guo-Qing Chang, Olga Karatayev, and Sarah F. Leibowitz, "Prenatal Exposure to Nicotine Stimulates Neurogenesis of Orexigenic Peptide-Expressing Neurons in Hypothalamus and Amygdala," *Journal of Neuroscience* 33(34) (2013): 13600–611.

22. Yann C. Klimentidis, T. Mark Beasley, Hui-Yi Lin, Giulianna Murati, Gregory E. Glass, Marcus Guyton, Wendy Newton, et al., "Canaries in the Coal Mine: A Cross-Species Analysis of the Plurality of Obesity Epidemics," *Proceedings of the Royal Society B* 278(1712) (2011): 1626–32.

23. Kristina A. Thayer, Jerrold J. Heindel, John R. Bucher, and Michael A. Gallo, "Role of Environmental Chemicals in Diabetes and Obesity: A National Toxicology Program Workshop Review," *Environmental Health Perspectives* 120(6) (2012): 779–89.

24. Séverine Kirchner, Tiffany Kieu, Connie Chow, Stephanie Casey, Bruce Blumberg, "Prenatal Exposure to the Environmental Obesogen Tributyltin Predisposes Multipotent Stem Cells to Become Adipocytes," *Molecular Endocrinology* 24(3) (2010): 526–39.

25. Frederick S. Vom Saal, Susan C. Nagel, Benjamin L. Coe, Brittany M. Angle, and Julia A. Taylor, "The Estrogenic Endocrine Disrupting Compound Bisphenol A (BPA) and Obesity," *Molecular and Cellular Endocrinology* 354(1–2) (2012): 74–84.

26. Bastiaan T. Heijmans, Elmar W. Tobi, Aryeh D. Stein, Hein Putter, Gerard J. Blauw, Ezra S. Susser, P. Eline Slagboom, and L. H. Lumey, "Persistent Epigenetic Differences Associated with Prenatal Exposure to Famine in Humans," *Proceedings of the National Academy of Sciences USA* 105(44) (2008): 17046–49.

27. Michael K. Skinner, Mohan Manikkam, Rebecca Tracey, Carlos Guerrero-Bosagna, Muksitul Haque, and Eric E. Nilsson, "Ancestral Dichlorodiphenyltrichloroethane (DDT) Exposure Promotes Epigenetic Transgenerational Inheritance of Obesity," *BMC Medicine* 11 (2013): 228.

28. Gregory A. Dunn and Tracy L. Bale, "Maternal High-Fat Diet Promotes

Body Length Increases and Insulin Insensitivity in Second-Generation Mice," *Endocrinology* 150(11) (2009): 4999–5009.

29. G. Kaati, L. O. Bygren, and S. Edvinsson, "Cardiovascular and Diabetes Mortality Determined by Nutrition During Parents' and Grandparents' Slow Growth Period," *European Journal of Human Genetics* 10(11) (2002): 682–88.

30. S. R. Veena, S. Geetha, S. D. Leary, J. Saperia, D. J. Fisher, K. Kumaran, P. Coakley, et al., "Relationships of Maternal and Paternal Birthweights to Features of the Metabolic Syndrome in Adult Offspring: An Inter-Generational Study in South India," *Diabetologia* 50(1) (2007): 43–54.

31. Ruth E. Brown, Arya M. Sharma, Chris I. Ardern, Pedi Mirdamadi, Paul Mirdamadi, and Jennifer L. Kuk, "Secular Differences Between Caloric Intake, Macronutrient Intake, and Physical Activity with Obesity," *Obesity Research & Clinical Practice* (2015), online ahead of print. DOI: 10.1016/j.orcp.2015.08.007.

Chapter 7: Stress, Shame, and Stigma

1. Bernardo Léo Wajchenberg, "Subcutaneous and Visceral Adipose Tissue: Their Relation to the Metabolic Syndrome," *Endocrine Reviews* 21(6) (2000): 697–738.

2. J. Marniemi, E. Kronholm, S. Aunola, T. Toikka, C.-E. Mattlar, M. Koskenvuo, and T. Rönnemaa, "Visceral Fat and Psychosocial Stress in Identical Twins Discordant for Obesity," *Journal of Internal Medicine* 251(1) (2002): 35–43.

3. Mary F. Dallman, Norman Pecoraro, Susan F. Akana, Susanne E. la Fleur, Francisca Gomez, Hani Houshyar, M. E. Bell, et al., "Chronic Stress and Obesity: A New View of 'Comfort Food,'" *Proceedings of the National Academy of Sciences USA* 100(20) (2003): 11696–701.

4. Ilia N. Karatsoreos, Sarah M. Bhagat, Nicole P. Bowles, Zachary M. Weil, Donald W. Pfaff, and Bruce S. McEwen, "Endocrine and Physiological Changes in Response to Chronic Corticosterone: A Potential Model of the Metabolic Syndrome in Mouse," *Endocrinology* 151(5) (2010): 2117–27.

5. Kellie L. K. Tamashiro, Maria A. Hegeman, Mary M. N. Nguyen, Susan J. Melhorn, Li Yun Ma, Stephen C. Woods, and Randall R. Sakai, "Dynamic

Body Weight and Body Composition Changes in Response to Subordination Stress," *Physiology & Behavior* 91(4) (2007): 440–48.

6. Rachel K. Crowley, Beverly A. Hughes, Joanna Gray, Theresa McCarthy, Susan Hughes, Cedric H. L. Shackleton, Nicola Crabtree, et al., "Longitudinal Changes in Glucocorticoid Metabolism Are Associated with Later Development of Adverse Metabolic Phenotype," *European Journal of Endocrinology* 171(4) (2014): 433–42.
7. M. Novak, L. Björck, K. W. Giang, C. Heden-Ståhl, L. Wilhelmsen, and A. Rosengren, "Perceived Stress and Incidence of Type 2 Diabetes: A 35-Year Follow-up Study of Middle-Aged Swedish Men," *Diabetes Medicine* 30(1) (2013): e816.
8. Stefan Branth, Gunnar Ronquist, Mats Stridsberg, Leif Hambraeus, Erik Kindgren, Roger Olsson, David Carlander, and Bengt Ametz, "Development of Abdominal Fat and Incipient Metabolic Syndrome in Young Healthy Men Exposed to Long-Term Stress," *Nutrition, Metabolism and Cardiovascular Diseases* 17(6) (2007): 427–35.
9. Lydia E. Kuo, Joanna B. Kitlinska, Jason U. Tilan, Lijun Li, Stephen B. Baker, Michael D. Johnson, Edward W. Lee, et al., "Neuropeptide Y Acts Directly in the Periphery on Fat Tissue and Mediates Stress-Induced Obesity and Metabolic Syndrome," *Nature Medicine* 13(7) (2007): 803–11.
10. A. Janet Tomiyama, "Weight Stigma Is Stressful: A Review of Evidence for the Cyclic Obesity/Weight-Based Stigma Model," *Appetite* 82 (2014): 8–15.
11. Alvin R. Tarlov, "Public Policy Frameworks for Improving Population Health," *Annals of the New York Academy of Sciences* 896 (1999): 281–93.
12. Taroni Chandola, Eric Brunner, and Michael Marmot, "Chronic Stress at Work and the Metabolic Syndrome: Prospective Study," *British Medical Journal,* 332(7540) (2006): 521–25.
13. Robert M. Sapolsky, *Why Zebras Don't Get Ulcers: The Acclaimed Guide to Stress, Stress-Related Diseases and Coping,* 3rd ed. (New York: W. H. Freeman, 2004).
14. Margaret E. Kemeny, Tara L. Gruenewald, and Sally S. Dickerson, "Shame as the Emotional Response to Threat to the Social Self: Implications for Behavior, Physiology, and Health," *Psychological Inquiry* 15(2) (2004): 153–60.

15. Vincent L. Wester, Sabine M. Staufenbiel, Margriet A. B. Veldhorst, Jenny A. Visser, Laura Manenschijn, Jan W. Koper, Francoise J. M. Klessens-Godfroy, et al., "Long-Term Cortisol Levels Measured in Scalp Hair of Obese Patients," *Obesity* 22(9) (2014): 1956–58.

16. Margriet Veldhorst, Gerard Noppe, Mieke H. T. M. Jongejan, Chantine B. M. Kok, Selma Mekic, Jan Willem Koper, Elisabeth F. C. van Rossum, and Erica L. T. van den Akker, "Increased Scalp Hair Cortisol Concentrations in Obese Children," *Journal of Clinical Endocrinology & Metabolism* 99(1) (2014): 285–90.

17. Natasha A. Schvey, Rebecca M. Puhl, and Kelly D. Brownell, "The Stress of Stigma: Exploring the Effect of Weight Stigma on Cortisol Reactivity," *Psychosomatic Medicine* 76(2) (2014): 156–62.

18. Rebecca M. Puhl and Chelsea A. Heuer, "Obesity Stigma: Important Considerations for Public Health," *American Journal of Public Health* 100(6) (2010): 1019–28.

19. Angelina R. Sutin, Yannick Stephan, Martina Luchetti, and Antonio Terracciano, "Perceived Weight Discrimination and C-Reactive Protein," *Obesity* 22(9) (2014): 1959–61.

20. A. Janet Tomiyama, Elissa Epel, Trissa A. McClatchey, Gina Poelke, Margaret E. Kemeny, Shannon K. McCoy, and Jennifer Daubenmier, "Associations of Weight Stigma with Cortisol and Oxidative Stress Independent of Adiposity," *Health Psychology* 33(8) (2014): 862–67.

21. S. J. Lupien, S. King, M. J. Meaney, and B. S. McEwen, "Can Poverty Get Under Your Skin? Basal Cortisol Levels and Cognitive Function in Children from Low and High Socioeconomic Status," *Developmental Psychopathology* 13(3) (2001): 653–76.

22. Anandi Mani, Sendhil Mullainathan, Eldar Shafir, and Jiaying Zhao, "Poverty Impedes Cognitive Function," *Science* 341(6149) (2013): 976–80.

23. Floriana S. Luppino, Leonore M. de Wit, Paul F. Bouvy, Theo Stijnen, Pim Cuijpers, Brenda W. J. H. Penninx, and Frans G. Zitman, "Overweight, Obesity, and Depression: A Systematic Review and Meta-analysis of Longitudinal Studies," *Archives of General Psychiatry* 67(3) (2010): 220–29.

24. S. R. Bornstein, A. Schuppenies, M.-L. Wong, and J. Licinio, "Approaching the Shared Biology of Obesity and Depression: The Stress Axis as the Locus

of Gene-Environment Interactions," *Molecular Psychiatry* 11(10) (2006): 892–902.

25. Mary F. Dallman, "Stress-Induced Obesity and the Emotional Nervous System," *Trends in Endocrinology and Metabolism* 21(3) (2010): 150–65.

26. A. Janet Tomiyama, Traci Mann, Danielle Vinas, Jeffrey M. Hunger, Jill DeJager, and Shelley E. Taylor, "Low Calorie Dieting Increases Cortisol," *Psychosomatic Medicine* 72(4) (2010): 357–64.

27. Drew A. Anderson, Jennifer R. Shapiro, Jennifer D. Lundgren, Leah E. Spataro, and Cheryl A. Frye, "Self-Reported Dietary Restraint Is Associated with Elevated Levels of Salivary Cortisol," *Appetite* 38(1) (2002): 13–17.

28. Judy A. McLean, Susan I. Barr, and Jerilynn C. Prior, "Cognitive Dietary Restraint Is Associated with Higher Urinary Cortisol Excretion in Healthy Premenopausal Women," *American Journal of Clinical Nutrition* 73(1) (2001): 7–12.

29. Amy Kiefer, Jue Lin, Elizabeth Blackburn, and Elissa Epel, "Dietary Restraint and Telomere Length in Pre- and Postmenopausal Women," *Psychosomatic Medicine* 70(8) (2008): 845–49.

30. Eli Puterman, Jue Lin, Elizabeth Blackburn, Aoife O'Donovan, Nancy Adler, and Elissa Epel, "The Power of Exercise: Buffering the Effect of Chronic Stress on Telomere Length," *PLoS ONE* 5(5) (2010): e10837.

31. Rajita Sinha and Ania M. Jastreboff, "Stress as a Common Risk Factor for Obesity and Addiction," *Biological Psychiatry* 73(9) (2013): 827–35.

32. Harriët Schellekens, Beate C. Finger, Timothy G. Dinan, and John F. Cryan, "Ghrelin Signalling and Obesity: At the Interface of Stress, Mood and Food Reward," *Pharmacology & Therapeutics* 135(3) (2012): 316–26.

33. P. A. Tataranni, D. E. Larson, S. Snitker, J. B. Young, J. P. Flatt, and E. Ravussin, "Effects of Glucocorticoids on Energy Metabolism and Food Intake in Humans," *American Journal of Physiology—Endocrinology and Metabolism* 271(2) (1996): E317–E325.

34. A. Janet Tomiyama, Mary F. Dallman, and Elissa S. Epel, "Comfort Food Is Comforting to Those Most Stressed: Evidence of the Chronic Stress Response Network in High Stress Women," *Psychoneuroendocrinology* 36(10) (2011): 1513–19.

35. Diana E. Pankevich, Sarah L.Teegarden, Andrew D. Hedin, Catherine L. Jensen, and Tracy L. Bale, "Calorie Restriction Experience Reprograms Stress and Orexigenic Pathways and Promotes Binge Eating," *Journal of Neuroscience* 30(48) (2010): 16399–407.

36. Eve Van Cauter and Kristen L. Knutson, "Sleep and the Epidemic of Obesity in Children and Adults," supplement, *European Journal of Endocrinology* 159(1) (2008): S59–S66.

37. L. S. Nielsen, K. V. Danielsen, and T. I. A. Sørensen, "Short Sleep Duration as a Possible Cause of Obesity: Critical Analysis of the Epidemiological Evidence," *Obesity Reviews* 12(2) (2011): 78–92.

38. Lorrie Magee and Lauren Hale, "Longitudinal Associations Between Sleep Duration and Subsequent Weight Gain: A Systematic Review," *Sleep Medicine Reviews* 16(3) (2012): 231–41.

39. Laura Manenschijn, Rulanda G. P. M. van Kruysbergen, Frank H. de Jong, Jan W. Koper, and Elisabeth F. C. van Rossum, "Shift Work at Young Age Is Associated with Elevated Long-Term Cortisol Levels and Body Mass Index," *Journal of Clinical Endocrinology & Metabolism* 96(11) (2011): E1862–E1865.

40. Kristen L. Knutson and Eve Van Cauter, "Associations Between Sleep Loss and Increased Risk of Obesity and Diabetes," *Annals of the New York Academy of Sciences* 1129 (2008): 287–304.

41. Francesco P. Cappuccio, Lanfranco D'Elia, Pasquale Strazzullo, and Michelle A. Miller, "Quantity and Quality of Sleep and Incidence of Type 2 Diabetes: A Systematic Review and Meta-analysis," *Diabetes Care* 33(2) (2010): 414–20.

42. Qun Xu, Yiqing Song, Albert Hollenbeck, Aaron Blair, Arthur Schatzkin, and Honglei Chen, "Day Napping and Short Night Sleeping Are Associated with Higher Risk of Diabetes in Older Adults," *Diabetes Care* 33(1) (2010): 78–83.

43. Paul Grossman, Ludger Niemann, Stefan Schmidt, and Harald Walach, "Mindfulness-Based Stress Reduction and Health Benefits: A Meta-analysis," *Journal of Psychosomatic Research* 57(1) (2004): 35–43.

44. Michael Gleeson, Nicolette C. Bishop, David J. Stensel, Martin R. Lindley, Sarabjit S. Mastana, and Myra R. Nimmo, "The Anti-inflammatory Effects

of Exercise: Mechanisms and Implications for the Prevention and Treatment of Disease," *Nature Reviews Immunology* 11(9) (2011): 607–15.

45. Jeffrey A. Woods, Kenneth R. Wilund, Stephen A. Martin, and Brandon M. Kistler, "Exercise, Inflammation and Aging," *Aging and Disease* 3(1) (2012): 130–40.

46. Chun-Xia Yi, Omar Al Massadi, Elizabeth Donelan, Maarit Lehti, Jon Weber, Chandler Ress, Chitrang Trivedi, et al., "Exercise Protects Against High-Fat Diet-Induced Hypothalamic Inflammation," *Physiology & Behavior* 106(4) (2012): 485–90.

Chapter 8: When Calories Don't Count

1. Lorien E. Urban, Gerard E. Dallial, Lisa M. Robinson, Lynne M. Ausman, Edward Saltzman, and Susan B. Roberts, "The Accuracy of Stated Energy Contents of Reduced-Energy, Commercially Prepared Foods," *Journal of the American Dietetic Association* 110(1) (2010): 116–23.

2. Kevin D. Hall, Steven B. Heymsfield, Joseph W. Kemnitz, Samuel Klein, Dale A. Schoeller, and John R. Speakman, "Energy Balance and Its Components: Implications for Body Weight Regulation," *American Journal of Clinical Nutrition* 95(4) (2012): 989–94.

3. Alexandra M. Johnstone, Sandra D. Murison, Jackie S. Duncan, Kellie A. Rance, and John R. Speakman, "Factors Influencing Variation in Basal Metabolic Rate Include Fat-Free Mass, Fat Mass, Age, and Circulating Thyroxine but Not Sex, Circulating Leptin, or Triiodothyronine," *American Journal of Clinical Nutrition* 82(5) (2005): 941–48.

4. Casey Neistat, "Calorie Detective," *New York Times,* February 12, 2013.

5. Urban, Dallial, Robinson, Ausman, Saltzman, and Roberts, "The Accuracy of Stated Energy Contents of Reduced-Energy, Commercially Prepared Foods."

6. Janet A. Novotny, Sarah K. Gebauer, and David J. Baer, "Discrepancy Between the Atwater Factor Predicted and Empirically Measured Energy Values of Almonds in Human Diets," *American Journal of Clinical Nutrition* 96(2) (2012): 296–301.

7. Rachel N. Carmody and Richard W. Wrangham, "The Energetic Significance of Cooking," *Journal of Human Evolution* 57(4) (2009): 379–91.

8. Karina Fonseca-Azevedo and Suzana Herculano-Houzel, "Metabolic Constraint Imposes Tradeoff Between Body Size and Number of Brain Neurons in Human Evolution," *Proceedings of the National Academy of Sciences USA* 109(45) (2012): 18571–76.

9. C. Koebnick, C. Strassner, I. Hoffmann, and C. Leitzmann, "Consequences of a Long-Term Raw Food Diet on Body Weight and Menstruation: Results of a Questionnaire Survey," *Annals of Nutrition & Metabolism* 43(2) (1999): 69–78.

10. Scott M. Boback, Christian L. Cox, Brian D. Ott, Rachel Carmody, Richard W. Wrangham, and Stephen M. Secor, "Cooking and Grinding Reduces the Cost of Meat Digestion," *Comparative Biochemistry and Physiology, Part A: Molecular & Integrative Physiology* 148(3) (2007): 651–56.

11. Pieter Evenepoel, Benny Geypens, Anja Luypaerts, Martin Hiele, Yvo Ghoos, and Paul Rutgeerts, "Digestibility of Cooked and Raw Egg Protein in Humans as Assessed by Stable Isotope Techniques," *Journal of Nutrition* 128(10) (1998): 1716–22.

12. Emily E. Groopman, Rachel N. Carmody, and Richard W. Wrangham, "Cooking Increases Net Energy Gain from a Lipid-Rich Food," *American Journal of Physical Anthropology* 156(1) (2015): 11–18.

13. Dariush Mozaffarian, Tao Hao, Eric B. Rimm, Walter C. Willett, and Frank B. Hu, "Changes in Diet and Lifestyle and Long-Term Weight Gain in Men and Women," *New England Journal of Medicine* 364(25) (2011): 2393–404.

14. Fredrik Bäckhed, Hao Ding, Ting Wang, Lora V. Hooper, Gou Young Koh, Andras Nagy, Clay F. Semenkovich, and Jeffrey I. Gordon, "The Gut Microbiota as an Environmental Factor That Regulates Fat Storage," *Proceedings of the National Academy of Sciences USA* 101(44) (2004): 15718–23.

15. Lawrence A. David, Corinne F. Maurice, Rachel N. Carmody, David B. Gootenberg, Julie E. Button, Benjamin E. Wolfe, Alisha V. Ling, et al., "Diet Rapidly and Reproducibly Alters the Human Gut Microbiome," *Nature* 505(7484) (2014): 559–63.

16. Martin J. Blaser and Stanley Falkow, "What Are the Consequences of the

Disappearing Human Microbiota?" *Nature Reviews Microbiology* 7(12) (2009): 887–94.

17. T. Azuma, H. Suto, Y. Ito, M. Ohtani, M. Dojo, M. Kuriyama, and T. Kato, "Gastric Leptin and *Helicobacter Pylori* Infection," *Gut* 49(3) (2001): 324–29.

18. Julia K. Goodrich, Jillian L. Waters, Angela C. Poole, Jessica L. Sutter, Omry Koren, Ran Blekhman, Michelle Beaumont, et al., "Human Genetics Shape the Gut Microbiome," *Cell* 159(4) (2014): 789–99.

19. Vanessa K. Ridaura, Jeremiah J. Faith, Federico E. Rey, Jiye Cheng, Alexis E. Duncan, Andrew L. Kau, Nicholas W. Griffin, et al., "Gut Microbiota from Twins Discordant for Obesity Modulate Metabolism in Mice," *Science* 341(6150) (2013). DOI: 10.1126/science.1241214.

20. Giovanni Musso, Roberto Gambino, and Maurizio Cassader, "Interactions Between Gut Microbiota and Host Metabolism Predisposing to Obesity and Diabetes," *Annual Review of Medicine* 62 (2011): 361–80.

21. Alice P. Liou, Melissa Paziuk, Jesus-Mario Luevano Jr., Sriram Machineni, Peter J. Turnbaugh, and Lee M. Kaplan, "Conserved Shifts in the Gut Microbiota Due to Gastric Bypass Reduce Host Weight and Adiposity," *Science Translational Medicine* 5(178) (2013): 178ra41.

22. Neha Alang and Colleen R. Kelly, "Weight Gain After Fecal Microbiota Transplantation," *Open Forum Infectious Diseases* 2(1) (2015). DOI: 10.1093/ofid/ofv004.

23. Karen K. Ryan, Valentina Tremaroli, Christoffer Clemmensen, Petia Kovatcheva-Datchary, Andriy Myronovych, Rebekah Karns, Hilary E. Wilson-Pérez, et al., "FXR Is a Molecular Target for the Effects of Vertical Sleeve Gastrectomy," *Nature* 509(7499) (2014): 183–88.

24. Sungsoon Fang, Jae Myoung Suh, Shannon M. Reilly, Elizabeth Yu, Olivia Osborn, Denise Lackey, Eiji Yoshihara, et al., "Intestinal FXR Agonism Promotes Adipose Tissue Browning and Reduces Obesity and Insulin Resistance," *Nature Medicine* 21(2) (2015): 159–65.

25. Nadja Larsen, Finn K. Vogensen, Frans W. J. van den Berg, Dennis Sandris Nielsen, Anne Sofie Andreassen, Bente K. Pedersen, Waleed Abu Al-Sou, et al., "Gut Microbiota in Human Adults with Type 2 Diabetes Differs from Nondiabetic Adults," *PLoS ONE* 5(2) (2010): e9085.

26. Anne Vrieze, Els Van Nood, Frits Holleman, Jarkko Salojärvi, Ruud S. Kootte, Joep F. W. M. Bartelsman, Geesje M. Dallinda-Thie, et al., "Transfer of Intestinal Microbiota from Lean Donors Increases Insulin Sensitivity in Individuals with Metabolic Syndrome," *Gastroenterology* 143(4) (2012): 913–16.

27. Na Fei and Liping Zhao, "An Opportunistic Pathogen Isolated from the Gut of an Obese Human Causes Obesity in Germfree Mice," *ISME Journal* 7(4) (2013): 880–84.

28. Patrice D. Cani, Jacques Amar, Miguel Angel Iglesias, Marjorie Poggi, Claude Knauf, Delphine Bastelica, Audrey M. Neyrinck, et al., "Metabolic Endotoxemia Initiates Obesity and Insulin Resistance," *Diabetes* 56(7) (2007): 1761–72.

29. Omry Koren, Julia K. Goodrich, Tyler C. Cullender, Aymé Spor, Kirsi Laitinen, Helene Kling Bäckhed, Antonio Gonzalez, et al., "Host Remodeling of the Gut Microbiome and Metabolic Changes During Pregnancy," *Cell* 150(3) (2012): 470–80.

30. Emmanuelle Le Chatelier, Trine Nielsen, Junjie Qin, Edi Prifti, Falk Hildebrand, Gwen Falony, Mathieu Almeida, et al., "Richness of Human Gut Microbiome Correlates with Metabolic Markers," *Nature* 500(7464) (2013): 541–46.

31. Ilseung Cho, Shingo Yamanishi, Laura Cox, Barbara A. Methé, Jiri Zavadil, Kelvin Li, Zhan Gao, et al., "Antibiotics in Early Life Alter the Murine Colonic Microbiome and Adiposity," *Nature* 488(7413) (2012): 621–26.

32. Laura M. Cox, Shingo Yamanishi, Jiho Sohn, Alexander V. Alekseyenko, Jacqueline M. Leung, Ilseung Cho, Sungheon G. Kim, et al., "Altering the Intestinal Microbiota During a Critical Developmental Window Has Lasting Metabolic Consequences," *Cell* 158(4) (2014): 705–21.

33. M. B. Azad, S. L. Bridgman, A. B. Becker, and A. L. Kozyrskyj, "Infant Antibiotic Exposure and the Development of Childhood Overweight and Central Adiposity," *International Journal of Obesity* 38(10) (2014): 1290–98.

34. T. A. Ajslev, C. S. Andersen, M. Gamborg, T. I. A. Sørensen, and T. Jess, "Childhood Overweight After Establishment of the Gut Microbiota: The

Role of Delivery Mode, Pre-pregnancy Weight and Early Administration of Antibiotics," *International Journal of Obesity* 35(4) (2011): 522–29.

35. Mike Sharland on behalf of the SACAR Paediatric Subgroup, "The Use of Antibacterials in Children: A Report of the Specialist Advisory Committee on Antimicrobial Resistance (SACAR) Paediatric Subgroup," supplement, *Journal of Antimicrobial Chemotherapy* 60(1) (2007): i15–i26.

36. Anders Hviid, Henrik Svanström, and Morten Frisch, "Antibiotic Use and Inflammatory Bowel Disease in Childhood," *Gut* 60 (2011): 49–54.

Chapter 9: Blame Your Ancestors

1. Lisa Belkin, "Watching Her Weight," *New York Times Magazine,* July 8, 2001.

2. Story Rick Nathanson, "Nine Years Ago, the State Took Custody of 3-Year-Old Ana Because of Her Weight; She Still Faces Challenges Today," *Albuquerque Journal,* February 28, 2010.

3. Daniel R. Gross, "Time Allocation: A Tool for the Study of Cultural Behavior," *Annual Review of Anthropology* 13 (1984): 519–58.

4. Anna Revedin, Biancamaria Aranguren, Roberto Becattini, Laura Longo, Emanuele Marconi, Marta Mariotti Lippi, Natalia Skakun, et al., "Thirty Thousand-Year-Old Evidence of Plant Food Processing," *Proceedings of the National Academy of Sciences USA* 107(44) (2010): 18815–19.

5. Frank W. Marlowe, "Hunter-Gatherers and Human Evolution," *Evolutionary Anthropology* 14(2) (2005): 54–67.

6. Kevin N. Laland, John Odling-Smee, and Sean Myles, "How Culture Shaped the Human Genome: Bringing Genetics and the Human Sciences Together," *Nature Reviews Genetics* 11(2) (2010): 137–48.

7. K. M. Flegal and R. P. Troiano, "Changes in the Distribution of Body Mass Index of Adults and Children in the US Population," *International Journal of Obesity and Related Metabolic Disorders* 24(7) (2000): 807–18.

8. Claudia Dreifus, "A Mathematical Challenge to Obesity: A Conversation with Carson Chow," *New York Times,* May 14, 2012.

9. B. Rokholm, J. L. Baker, and T. Sørensen, "The Levelling Off of the Obesity Epidemic Since the Year 1999—A Review of Evidence and Perspectives," *Obesity Reviews* 11(12) (2010): 835–46.

10. K. Silventoinen, B. Rokholm, J. Kaprio, and T. I. A. Sørensen, "The Genetic and Environmental Influences on Childhood Obesity: A Systematic Review of Twin and Adoption Studies," *International Journal of Obesity* 34(1) (2010): 29–40.

11. John J. Reilly, Julie Armstrong, Ahmad R. Dorosty, Pauline M. Emmett, A. Ness, I. Rogers, Colin Steer, et al., "Early Life Risk Factors for Obesity in Childhood: Cohort Study," *British Medical Journal* 330 (2005): 1357.

12. Jane Wardle, Susan Carnell, Claire M. A. Haworth, and Robert Plomin, "Evidence for a Strong Genetic Influence on Childhood Adiposity Despite the Force of the Obesogenic Environment," *American Journal of Clinical Nutrition* 87(2) (2008): 398–404.

13. J. B. Meigs, L. A. Cupples, and P. W. Wilson, "Parental Transmission of Type 2 Diabetes Mellitus: The Framingham Offspring Study," *Diabetes* 49(12) (2000): 2201–7.

14. Pia Skov Hansen, Thomas Heiberg Brix, Thorkild I. A. Sørensen, Kirsten Ohm Kyvik, and Laszlo Hegedüs, "Major Genetic Influence on the Regulation of the Pituitary-Thyroid Axis: A Study of Healthy Danish Twins," *Journal of Clinical Endocrinology & Metabolism* 89(3) (2004): 1181–87.

15. Mark P. J. Vanderpump, "The Epidemiology of Thyroid Disease," *British Medical Bulletin* 99(1) (2011): 39–51.

16. Nils Knudsen, Peter Laurberg, Lone B. Rasmussen, Inge Bülow, Hans Perrild, Lars Ovesen, and Torben Jørgensen, "Small Differences in Thyroid Function May Be Important for Body Mass Index and the Occurrence of Obesity in the Population," *Journal of Clinical Endocrinology & Metabolism* 90(7) (2005): 4019–24.

17. Carl T. Montague, I. Sadaf Farooqi, Jonathan P. Whitehead, Maria A. Soos, Harald Rau, Nicholas J. Wareham, Ciaran P. Sewter, et al., "Congenital Leptin Deficiency Is Associated with Severe Early-Onset Obesity in Humans," *Nature* 387(6636) (1997): 903–8.

18. I. Sadaf Farooqi, Susan A. Jebb, Gill Langmack, Elizabeth Lawrence, Christopher H. Cheetham, Andrew M. Prentice, Ieuan A. Hughes, et al.,

"Effects of Recombinant Leptin Therapy in a Child with Congenital Leptin Deficiency," *New England Journal of Medicine* 341(12) (1999): 879–84.

19. Oliver J. Muensterer and Riaz A. Agha, "Laparoscopic Sleeve Gastrectomy for a Two and a Half Year Old Morbidly Obese Child—A Leap into the Unknown," *International Journal of Surgery Case Reports* 4(11) (2013): 1055–56.
20. Heiko Krude, Heike Biebermann, Werner Luck, Rüdiger Horn, Georg Brabant, and Annette Grüters, "Severe Early-Onset Obesity, Adrenal Insufficiency and Red Hair Pigmentation Caused by POMC Mutations in Humans," *Nature Genetics* 19(2) (1998): 155–57.
21. Stephen O'Rahilly and I. Sadaf Farooqi, "Human Obesity: A Heritable Neurobehavioral Disorder That Is Highly Sensitive to Environmental Conditions," *Diabetes* 57(11) (2008): 2905–10.
22. Khalid K. Alharbi, Emmanuel Spanakis, Karen Tan, Matt J. Smith, Mohammed A. Aldahmesh, Sandra D. O'Dell, Avan Aihie Sayer, et al., "Prevalence and Functionality of Paucimorphic and Private MC4R Mutations in a Large, Unselected European British Population, Scanned by MeltMADGE," *Human Mutation* 28(3) (2007): 294–302.
23. Laura R. Pearce, Neli Atanassova, Matthew C. Banton, Bill Bottomley, Agatha A. van der Klaauw, Jean-Pierre Revelli, Audrey Hendricks, et al., "KSR2 Mutations Are Associated with Obesity, Insulin Resistance, and Impaired Cellular Fuel Oxidation," *Cell* 155(4) (2013): 765–77.
24. Timothy M. Frayling, Nicholas J. Timpson, Michael N. Weedon, Eleftheria Zeggini, Rachel M. Freathy, Cecilia M. Lindgren, John R. B. Perry, et al., "A Common Variant in the FTO Gene Is Associated with Body Mass Index and Predisposes to Childhood and Adult Obesity," *Science* 316(5826) (2007): 889–94.
25. Scott Smemo, Juan J. Tena, Kyoung-Han Kim, Eric R. Gamazon, Noboru J. Sakabe, Carlos Gómez-Marin, Ivy Aneas, et al., "Obesity-Associated Variants Within *FTO* Form Long-Range Functional Connections with IRX3," *Nature* 507(7492) (2014): 371–75.
26. Leticia A. Brondani, Tais S. Assmann, Bianca M. de Souza, Ana P. Bouças, Luis H. Canani, and Daisy Crispim, "Meta-analysis Reveals the Association of Common Variants in the *Uncoupling Protein* (UCP) *1–3* Genes with Body Mass Index Variability," *PLoS ONE* 9(5) (2014): e96411.

27. R. W. Morgan, J. S. Speakman, and S. E. Grimshaw, "Inuit Myopia: An Environmentally Induced 'Epidemic'?" *Canadian Medical Association Journal* 112(5) (1975): 575–77.

28. Kathryn Rose, Wayne Smith, Ian Morgan, and Paul Mitchell, "The Increasing Prevalence of Myopia: Implications for Australia," *Clinical & Experimental Ophthalmology* 29(3) (2001): 116–20.

29. Kathryn A. Rose, Ian G. Morgan, Wayne Smith, George Burlutsky, Paul Mitchell, and Seang-Mei Saw, "Myopia, Lifestyle, and Schooling in Students of Chinese Ethnicity in Singapore and Sydney," *Archives of Ophthalmology* 126(4) (2008): 527–30.

30. Lisa A. Jones, Loraine T. Sinnott, Donald O. Mutti, Gladys L. Mitchell, Melvin L. Moeschberger, and Karla Zadnik, "Parental History of Myopia, Sports and Outdoor Activities, and Future Myopia," *Investigative Ophthalmology & Visual Science* 48(8) (2007): 3524–32.

31. Regan Ashby, Arne Ohlendorf, and Frank Schaeffel, "The Effect of Ambient Illuminance on the Development of Deprivation Myopia in Chicks," *Investigative Ophthalmology and Visual Science* 50(11) (2009): 5348–54.

32. Kathryn A. Rose, Ian G. Morgan, Wayne Smith, and Paul Mitchell, "High Heritability of Myopia Does Not Preclude Rapid Changes in Prevalence," *Clinical & Experimental Ophthalmology* 30(3) (2002): 168–72.

33. Atsuhiko Ichimura, Akira Hirasawa, Odile Poulain-Godefroy, Amélie Bonnefond, Takafumi Hara, Loïc Yengo, Ikuo Kimura, et al., "Dysfunction of Lipid Sensor GPR120 Leads to Obesity in Both Mouse and Human," *Nature* 483(7389) (2012): 350–54.

34. Claude Bouchard, France T. Dionne, Jean-Aimé Simoneau, and Marcel R. Boulay, "Genetics of Aerobic and Anaerobic Performances," *Exercise & Sport Sciences Reviews* 20(1) (1992): 27–58.

35. Claude Bouchard and Angelo Tremblay, "Genetic Influences on the Response of Body Fat and Fat Distribution to Positive and Negative Energy Balances in Human Identical Twins," supplement, *Journal of Nutrition* 127(5) (1997): 943S–947S.

36. M. A. Sarzynski, P. Jacobson, T. Rankinen, B. Carlsson, L. Sjöström,

C. Bouchard, and L. M. S. Carlsson, "Associations of Markers in 11 Obesity Candidate Genes with Maximal Weight Loss and Weight Regain in the SOS Bariatric Surgery Cases," *International Journal of Obesity* 35(5) (2011): 676–83.

37. Christopher D. Still, G. Craig Wood, Xin Chu, Robert Erdman, Christina H. Manney, Peter N. Benotti, Anthony T. Petrick, et al., "High Allelic Burden of Four Obesity SNPs Is Associated with Poorer Weight Loss Outcomes Following Gastric Bypass Surgery," *Obesity* 19(8) (2011): 1676–83.

38. H. R. Lajunen, J. Kaprio, A. Keski-Rahkonen, R. J. Rose, L. Pulkkinen, A. Rissanen, and K. Silventoinen, "Genetic and Environmental Effects on Body Mass Index During Adolescence: A Prospective Study Among Finnish Twins," *International Journal of Obesity* 33(5) (2009): 559–67.

39. Lise Dubois, Kirsten Ohm Kyvik, Manon Girard, Fabiola Tatone-Tokuda, Daniel Pérusse, Jacob Hjelmborg, Axel Skytthe, et al., "Genetic and Environmental Contributions to Weight, Height, and BMI from Birth to 19 Years of Age: An International Study of Over 12,000 Twin Pairs," *PLoS ONE* 7(2) (2012): e30153.

40. Sanna Tholin, Finn Rasmussen, Per Tynelius, and Jan Karlsson, "Genetic and Environmental Influences on Eating Behavior: The Swedish Young Male Twins Study," *American Journal of Clinical Nutrition* 81(3) (2005): 564–69.

41. J. Wardle, C. Llewellyn, S. Sanderson, and R. Plomin, "The FTO Gene and Measured Food Intake in Children," *International Journal of Obesity* 33(1) (2009): 42–45.

42. William S. Garver, Sara B. Newman, Diana M. Gonzales-Pacheco, Joseph J. Castillo, David Jelinek, Randall A. Heidenreich, and Robert A. Orlando, "The Genetics of Childhood Obesity and Interaction with Dietary Macronutrients," *Genes & Nutrition* 8(3) (2013): 271–87.

43. James Niels Rosenquist, Steven F. Lehrer, A. James O'Malley, Alan M. Zaslavsky, Jordan W. Smoller, and Nicholas Christakis, "Cohort of Birth Modifies the Association Between FTO Genotype and BMI," *Proceedings of the National Academy of Sciences USA* 112(2) (2015): 354–59.

Chapter 10: Follow the Money

1. Lisa Held, "Psychoanalysis Shapes Consumer Culture," *Monitor on Psychology* 40(11) (2009): 32.

2. Federal Trade Commission, "A Review of Food Marketing to Children and Adolescents: Follow-up Report," December 2012.

3. Cees de Graaf and Frans J. Kok, "Slow Food, Fast Food and the Control of Food Intake," *Nature Reviews Endocrinology* 6(5) (2010): 290–93.

4. D. P. DiMeglio and R. D. Mattes, "Liquid Versus Solid Carbohydrate: Effects on Food Intake and Body Weight," *International Journal of Obesity* 24(6) (2000): 794–800.

5. Terry L. Davidson and Susan E. Swithers, "Food Viscosity Influences Caloric Intake Compensation and Body Weight in Rats," *Obesity Research* 13(3) (2005): 537–44.

6. Benoit Chassaing, Omry Koren, Julie K. Goodrich, Angela C. Poole, Shanthi Srinivasan, Ruth E. Ley, and Andrew T. Gewirtz, "Dietary Emulsifiers Impact the Mouse Gut Microbiota Promoting Colitis and Metabolic Syndrome," *Nature* 519(7541) (2015): 92–96.

7. Frederick J. Zimmerman, "Using Marketing Muscle to Sell Fat: The Rise of Obesity in the Modern Economy," *Annual Review of Public Health* 32 (2011): 285–306.

8. Frederick J. Zimmerman and Janice F. Bell, "Associations of Television Content Type and Obesity in Children," *American Journal of Public Health* 100(2) (2010): 334–40.

9. Karen Glanz, Michael Basil, Edward Maibach, Jeanne Goldberg, and Dan Snyder, "Why Americans Eat What They Do: Taste, Nutrition, Cost, Convenience, and Weight Control Concerns as Influences on Food Consumption," *Journal of the American Dietetic Association* 98(10) (1998): 1118–26.

10. Mark A. Pereira, Alex I. Kartashov, Cara B. Ebbeling, Linda Van Horn, Martha L. Slattery, David R. Jacobs Jr., and David S. Ludwig, "Fast-Food Habits, Weight Gain, and Insulin Resistance (the CARDIA study): 15-year Prospective Analysis," *Lancet* 365(9453) (2005): 36–42.

11. P. Rozin, C. Fischler, S. Imada, A. Sarubin, and A. Wrzesniewski, "Attitudes to Food and the Role of Food in Life in the U.S.A., Japan, Flemish Belgium

and France: Possible Implications for the Diet-Health Debate," *Appetite* 33(2) (1999): 163–80.

12. Brad Plumer, "The $956 Billion Farm Bill in One Graph," *Washington Post,* January 28, 2014.

13. Lindsay P. Smith, Shu Wen Ng, and Barry M. Popkin, "Trends in US Home Food Preparation and Consumption: Analysis of National Nutrition Surveys and Time Use Studies from 1965–1966 to 2007–2008," *Nutrition Journal* 12 (2013): 45.

14. Chase Purdy, "Attack on Meat Has Industry Seeing Red," *Politico,* February 19, 2015.

15. Lenard Lesser, Cara B. Ebbeling, Merrill Goozner, David Wypij, and David S. Ludwig, "Relationship Between Funding Source and Conclusion Among Nutrition-Related Scientific Articles," *PLoS Medicine* 4 (2007): e5.

16. D. C. Wilks, H. Besson, A. K. Lindroos, and U. Ekelund, "Objectively Measured Physical Activity and Obesity Prevention in Children, Adolescents and Adults: A Systematic Review of Prospective Studies," *Obesity Reviews* 12(5) (2011): e119–e129.

17. Barbara J. Rolls, Dianne Engell, and Leann L. Birch, "Serving Portion Size Influences 5-Year-Old but Not 3-Year-Old Children's Food Intakes," *Journal of the American Dietetic Association* 100(2) (2000): 232–34.

18. J. Hartmann-Boyce, D. J. Johns, S. A. Jebb, P. Aveyard, and on behalf of the Behavioural Weight Management Review Group, "Effect of Behavioural Techniques and Delivery Mode on Effectiveness of Weight Management: Systematic Review, Meta-analysis and Meta-regression," *Obesity Reviews* 15(7) (2014): 598–609.

19. Richard L. Cleland, Walter C. Gross, Laura D. Koss, Matthew Daynard, and Karen M. Muoio, "Weight-Loss Advertising: An Analysis of Current Trends," Federal Trade Commission, September 2002.

20. Coulter Jones, John Fauber, and Kristina Fiore, "Diet Drug Companies Spend $60 Million Since 2010 amid Push to Win FDA Approval for New Products," *Milwaukee Journal Sentinel,* April 19, 2015.

21. Paul R. Thomas, ed., *Weighing the Options: Criteria for Evaluating Weight-Management Programs* (Washington, DC: National Academy Press, 1995).

Chapter 11: Eat with Attention and Joy

1. Jason M. Lavender, Bianca F. Jardin, and Drew A. Anderson, "Bulimic Symptoms in Undergraduate Men and Women: Contributions of Mindfulness and Thought Suppression," *Eating Behaviors* 10(4) (2009): 228–31.

2. Akihiko Masuda, Matthew Price, and Robert D. Latzman, "Mindfulness Moderates the Relationship Between Disordered Eating Cognitions and Disordered Eating Behaviors in a Non-clinical College Sample," *Journal of Psychopathology and Behavioral Assessment* 34(1) (2012): 107–15.

3. Eric B. Loucks, Willoughby B. Britton, Chanelle J. Howe, Charles B. Eaton, and Stephen L. Buka, "Positive Associations of Dispositional Mindfulness with Cardiovascular Health: The New England Family Study," *International Journal of Behavioral Medicine* 22(4) (2015): 540–50.

4. Holley S. Hodgins and Kathryn C. Adair, "Attentional Processes and Meditation," *Consciousness and Cognition* 19 (2010): 872–78.

5. Stefan G. Hofmann, Alice T. Sawyer, Ashley A. Witt, and Diana Oh, "The Effect of Mindfulness-Based Therapy on Anxiety and Depression: A Meta-analytic Review," *Journal of Consulting and Clinical Psychology* 78(2) (2010): 169–83.

6. Jon Vøllestad, Morten Birkeland Nielsen, and Geir Høstmark Nielsen, "Mindfulness- and Acceptance-Based Interventions for Anxiety Disorders: A Systematic Review and Meta-analysis," *British Journal of Clinical Psychology* 51(3) (2012): 239–60.

7. Jacob Piet and Esben Hougaard, "The Effect of Mindfulness-Based Cognitive Therapy for Prevention of Relapse in Recurrent Major Depressive Disorder: A Systematic Review and Meta-analysis," *Clinical Psychology Review* 31(6) (2011): 1032–40.

8. J. David Creswell, Baldwin M. Way, Naomi I. Eisenberger, and Matthew D. Lieberman, "Neural Correlates of Dispositional Mindfulness During Affect Labeling," *Psychosomatic Medicine* 69(6) (2007): 560–65.

9. L. O. Fjorback, M. Arendt, E. Ørnbøl, P. Fink, and H. Walach, "Mindfulness-Based Stress Reduction and Mindfulness-Based Cognitive Therapy—A Systematic Review of Randomized Controlled Trials," *Acta Psychiatrica Scandinavica* 124(2) (2011): 102–19.

10. Jonathan S. Abramowitz, David F. Tolin, and Gordon P. Street, "Paradoxical Effects of Thought Suppression: A Meta-analysis of Controlled Studies," *Clinical Psychology Review* 21(5) (2001): 683–703.

11. Hugo J. E. M. Alberts, Francine Schneider, and Carolien Martijn, "Dealing Efficiently with Emotions: Acceptance-Based Coping with Negative Emotions Requires Fewer Resources Than Suppression," *Cognition & Emotion* 26(5) (2012): 863–70.

12. Hugo J. E. M. Alberts, Sandra Mulkens, Maud Smeets, and Roy Thewissen, "Coping with Food Cravings: Investigating the Potential of a Mindfulness-Based Intervention," *Appetite* 55(1) (2010): 160–63.

13. Sally A. Dockendorff, Trent A. Petrie, Christy A. Greenleaf, and Scott Martin, "Intuitive Eating Scale: An Examination Among Early Adolescents," *Journal of Counseling Psychology* 59(4) (2012): 604–11.

14. TeriSue Smith and Steven R. Hawks, "Intuitive Eating, Diet Composition, and the Meaning of Food in Healthy Weight Promotion," *American Journal of Health Education* 37(3) (2006): 130–36.

15. Nina Van Dyke and Eric J. Drinkwater, "Relationships Between Intuitive Eating and Health Indicators: Literature Review," *Public Health Nutrition* 17(8) (2014): 1757–66.

16. Clara E. L. Madden, Sook Ling Leong, Andrew Gray, and Caroline C. Horwath, "Eating in Response to Hunger and Satiety Signals Is Related to BMI in a Nationwide Sample of 1601 Mid-age New Zealand Women," *Public Health Nutrition* 15(12) (2012): 2272–79.

17. Julie Gast, Amy Campbell Nielson, Anne Hunt, and Jason L. Leiker, "Intuitive Eating: Associations with Physical Activity Motivation and BMI," *American Journal of Health Promotion* 29(3) (2015): e91–e99.

18. Steven Hawks, Hala Madanat, Jaylyn Hawks, and Ashley Harris, "The Relationship Between Intuitive Eating and Health Indicators Among College Women," *American Journal of Health Education* 36(6) (2005): 331–36.

19. Tracy L. Tylka, "Development and Psychometric Evaluation of a Measure of Intuitive Eating," *Journal of Counseling Psychology* 53(2) (2006): 226–40.

20. Van Dyke and Drinkwater, "Relationships Between Intuitive Eating and Health Indicators."

21. Beate M. Herbert, Jens Blechert, Martin Hautzinger, Ellen Matthias, and Cornelia Herbert, "Intuitive Eating Is Associated with Interoceptive Sensitivity: Effects on Body Mass Index," *Appetite* 70 (2013): 22–30.

22. Beate M. Herbert, Eric R. Muth, Olga Pollatos, and Cornelia Herbert, "Interoception Across Modalities: On the Relationship Between Cardiac Awareness and the Sensitivity for Gastric Functions," *PLoS ONE* 7(5) (2012): e36646.

23. Ruth A. Baer, Sarah Fischer, and Debra B. Huss, "Mindfulness and Acceptance in the Treatment of Disordered Eating," *Journal of Rational-Emotive and Cognitive-Behavior Therapy* 23(4) (2005): 281–300.

24. Jean Kristeller, Ruth Q. Wolever, and Virgil Sheets, "Mindfulness-Based Eating Awareness Training (MB-EAT) for Binge Eating: A Randomized Clinical Trial," *Mindfulness* 5(3) (2014): 282–97.

25. Linda W. Craighead, *The Appetite Awareness Workbook: How to Listen to Your Body and Overcome Bingeing, Overeating, and Obsession with Food* (Oakland, CA: New Harbinger Publications, 2006).

Chapter 12: Sleepwalking Through Dinner

1. Brian Wansink, *Mindless Eating: Why We Eat More Than We Think* (New York: Bantam, 2010).

2. B. Wansink and S. B. Park, "At the Movies: How External Cues and Perceived Taste Impact Consumption Volume," *Food Quality and Preference* 12(1) (2001): 69–74.

3. David J. Mela, "Determinants of Food Choice: Relationships with Obesity and Weight Control," supplement, *Obesity Research* 9(S11) (2001): 249S–255S.

4. Julian Parfitt, Mark Barthel, and Sarah Macnaughton, "Food Waste Within Food Supply Chains: Quantification and Potential for Change to 2050," *Philosophical Transactions of the Royal Society B* 365(1554) (2010): 3065–81.

5. Brian Wansink, James E. Painter, and Jill North, "Bottomless Bowls: Why Visual Cues of Portion Size May Influence Intake," *Obesity Research* 13(1) (2005): 93–100.

6. Barbara E. Kahn and Brian Wansink, "The Influence of Assortment Structure on Perceived Variety and Consumption Quantities," *Journal of Consumer Research* 30(4) (2004): 519–33.

7. Jeffrey Sobal and Brian Wansink, "Kitchenscapes, Tablescapes, Platescapes, and Foodscapes: Influences of Microscale Built Environments on Food Intake," *Environment and Behavior* 39(1) (2007): 124–42.

8. Brian Wansink, "Environmental Factors That Increase the Food Intake and Consumption Volume of Unknowing Consumers," *Annual Review of Nutrition* 24 (2004): 455–79.

9. Barbara J. Rolls, Dianne Engell, and Leann L. Birch, "Serving Portion Size Influences 5-Year-Old but Not 3-Year-Old Children's Food Intakes," *Journal of the American Dietetic Association* 100(2) (2000): 232–34.

10. Leann L. Birch, Susan L. Johnson, Graciella Andresen, John C. Peters, and Marcia C. Schulte, "The Variability of Young Children's Energy Intake," *New England Journal of Medicine* 324(4) (1991): 232–35.

11. Katja Kröller, Dörte Jahnke, and Petra Warschburger, "Are Maternal Weight, Eating and Feeding Practices Associated with Emotional Eating in Childhood?" *Appetite* 65 (2013): 25–30.

12. Jennifer Orlet Fisher and Leann Lipps Birch, "Restricting Access to Palatable Foods Affects Children's Behavioral Response, Food Selection, and Intake," *American Journal of Clinical Nutrition* 69(6) (1999): 1264–72.

13. Leann L. Birch, Jennifer Orlet Fisher, and Kirsten Krahnstoever Davison, "Learning to Overeat: Maternal Use of Restrictive Feeding Practices Promotes Girls' Eating in the Absence of Hunger," *American Journal of Clinical Nutrition* 78(2) (2003): 215–20.

14. Jennifer Orlet Fisher and Leann L. Birch, "Eating in the Absence of Hunger and Overweight in Girls at 5 and 7 Y of Age," *American Journal of Clinical Nutrition* 76(1) (2002): 226–31.

15. Tanja V. E. Kral, David B. Allison, Leann L. Birch, Virginia A. Stallings, Reneé H. Moore, and Myles S. Faith, "Caloric Compensation and Eating in

the Absence of Hunger in 5- to 12-Y-Old Weight-Discordant Siblings," *American Journal of Clinical Nutrition* 96(3) (2012): 574–83.

16. Gina L. Tripicchio, Kathleen L. Keller, Cassandra Johnson, Angelo Pietrobelli, Moonseong Heo, and Myles S. Faith, "Differential Maternal Feeding Practices, Eating Self-Regulation, and Adiposity in Young Twins," *Pediatrics* 134(5) (2014): e1399–e1404.

17. Julie A. Mennella and Gary K. Beauchamp, "Flavor Experiences During Formula Feeding Are Related to Preferences During Childhood," *Early Human Development* 68(2) (2002): 71–82.

18. Ellyn Satter, *Child of Mine: Feeding with Love and Good Sense* (Boulder, CO: Bull Publishing, 2000).

19. Katja Kröller and Petra Warschburger, "Associations Between Maternal Feeding Style and Food Intake of Children with a Higher Risk for Overweight," *Appetite* 51(1) (2008): 166–72.

20. Elizabeth D. Capaldi, ed., *Why We Eat What We Eat: The Psychology of Eating* (Washington, DC: American Psychological Association, 1996), chapter 3.

21. Maureen K. Spill, Leann L. Birch, Liane S. Roe, and Barbara J. Rolls, "Hiding Vegetables to Reduce Energy Density: An Effective Strategy to Increase Children's Vegetable Intake and Reduce Energy Intake," *American Journal of Clinical Nutrition* 94(3) (2011): 735–41.

Chapter 13: Healthy Is Better Than Thin

1. Emma Lewis, "Why There's No Point Telling Me to Lose Weight," *British Medical Journal* 350 (2015): g6845.

2. Christine Aramburu, Alegria Drury, and Margaret Louis, "Exploring the Association Between Body Weight, Stigma of Obesity, and Health Care Avoidance," *Journal of the American Association of Nurse Practitioners* 14(12) (2002): 554–61.

3. Steven N. Blair, "Physical Inactivity: The Biggest Public Health Problem of the 21st Century," *British Journal of Sports Medicine* 43(1) (2009): 1–2.

4. dish lab, www.dishlab.org/hero.php.

5. Joyce B. Harp and Lindsay Hecht, "Obesity in the National Football League," *Journal of the American Medical Association* 293(9) (2005): 1061–62.

6. Katherine M. Flegal, Brian K. Kit, Heather Orpana, and Barry I. Graubard, "Association of All-Cause Mortality with Overweight and Obesity Using Standard Body Mass Index Categories: A Systematic Review and Meta-analysis," *Journal of the American Medical Association* 309(1) (2013): 71–82.

7. Virginia Hughes, "The Big Fat Truth," *Nature* 497(7450) (2013): 428–30.

8. Robert J. Kuczmarski and Katherine M. Flegal, "Criteria for Definition of Overweight in Transition: Background and Recommendations for the United States," *American Journal of Clinical Nutrition* 72(5) (2000): 1074–S81.

9. R. P. Troiano and K. M. Flegal, "Overweight Prevalence Among Youth in the United States: Why So Many Different Numbers?" supplement, *International Journal of Obesity and Related Metabolic Disorders* 23(2) (1999): S22–S27.

10. Yoni Freedhoff, "Nothing Good Can Ever Come from Weighing Your Child," *U.S. News & World Report,* November 27, 2013.

11. Stephanie E. Simonds, Jack T. Pryor, Eric Ravussin, Frank L. Greenway, Ralph Dileone, Andrew M. Allen, Jaspreet Bassi, et al., "Leptin Mediates the Increase in Blood Pressure Associated with Obesity," *Cell* 159(6) (2014): 1404–16.

12. Look AHEAD Research Group, "Cardiovascular Effects of Intensive Lifestyle Intervention in Type 2 Diabetes," *New England Journal of Medicine* 369(2) (2013): 145–54.

13. Amy Berrington de Gonzalez, Patricia Hartge, James R. Cerhan, Alan J. Flint, Lindsay Hannan, Robert J. MacInnis, Steven C. Moore, et al., "Body-Mass Index and Mortality Among 1.46 Million White Adults," *New England Journal of Medicine* 363 (2010): 2211–19.

14. Thorkild I. A. Sörensen, Aila Rissanen, Maarit Korkeila, and Jaakko Kaprio, "Intention to Lose Weight, Weight Changes, and 18-y Mortality in Overweight Individuals Without Co-morbidities," *PLoS Medicine* 2 (2005): e171.

15. A. G. Dulloo, J. Jacquet, J.-P. Montani, and Y. Schutz, "How Dieting Makes the Lean Fatter: From a Perspective of Body Composition Autoregulation

Through Adipostats and Proteinstats Awaiting Discovery," supplement, *Obesity Reviews* 16(S1) (2015): 25–35.

16. I-Min Lee and Ralph S. Paffenbarger Jr., "Change in Body Weight and Longevity," *Journal of the American Medical Association* 268(15) (1992): 2045–49.

17. Peter Rzehak, Christa Meisinger, Gabriele Woelke, Sabine Brasche, Gert Strube, and Joachim Heinrich, "Weight Change, Weight Cycling and Mortality in the ERFORT Male Cohort Study," *European Journal of Epidemiology* 22(10) (2007): 665–73.

18. Vanessa A. Diaz, Arch G. Mainous, and Charles J. Everett, "The Association Between Weight Fluctuation and Mortality: Results from a Population-Based Cohort Study," *Journal of Community Health* 30(3) (2005): 153–65.

19. Lauren Lissner, L. Patricia M. Odell, Ralph B. D'Agostino, Joseph Stokes, Bernard E. Kreger, Albert J. Belanger, and Kelly D. Brownell, "Variability of Body Weight and Health Outcomes in the Framingham Population," *New England Journal of Medicine* 324(26) (1991): 1839–44.

20. Victoria L. Stevens, Eric J. Jacobs, Juzhong Sun, Alpa V. Patel, Marjorie L. McCullough, Lauren R. Teras, and Susan M. Gapstur, "Weight Cycling and Mortality in a Large Prospective US Study," *American Journal of Epidemiology* 175(8) (2012): 785–92.

21. Alison E. Field, Susan Malspeis, and Walter C. Willett, "Weight Cycling and Mortality Among Middle-Aged or Older Women," *Archives of Internal Medicine* 169(9) (2009): 881–86.

22. A. Bosy-Westphal, J. Kahlhöfer, M. Lagerpusch, T. Skurk, and M. J. Muller, "Deep Body Composition Phenotyping During Weight Cycling: Relevance to Metabolic Efficiency and Metabolic Risk," supplement, *Obesity Reviews* 16(S1) (2015): 36–44.

23. A. G. Dulloo, J. Jacquet, J. Seydoux, and J.-P. Montani, "The Thrifty 'Catch-Up Fat' Phenotype: Its Impact on Insulin Sensitivity During Growth Trajectories to Obesity and Metabolic Syndrome," supplement, *International Journal of Obesity* 30(S4) (2006): S23–S35.

24. H. Yatsuya, K. Tamakoshi, T. Yoshida, Y. Hori, H. Zhang, M. Ishikawa, S. Zhu, et al., "Association Between Weight Fluctuation and Fasting Insulin

Concentration in Japanese Men," *International Journal of Obesity* 27(4) (2003): 478–83.

25. Huiming Zhang, Koji Tamakoshi, Hiroshi Yatsuya, Chlyoe Murata, Keiko Wada, Rei Otsuka, Nobue Nagasawa, et al., "Long-Term Body Weight Fluctuation Is Associated with Metabolic Syndrome Independent of Current Body Mass Index Among Japanese Men," *Circulation Journal* 69(1) (2005): 13–18.

26. Taeko Kajioka, Shigeki Tsuzuku, Hiroshi Shimokata, and Yuzo Sato, "Effects of Intentional Weight Cycling on Non-obese Young Women," *Metabolism* 51(2) (2002): 149–54.

27. Jason Andrade and Andrew Igneszewski, "Exercise and the Heart: A Review of the Early Studies, in Memory of Dr. R. S. Paffenbarger," *BC Medical Journal* 49(10) (2007): 540–46.

28. Vaughn W. Barry, Meghan Baruth, Michael W. Beets, J. Larry Durstine, Jihong Liu, and Steven N. Blair, "Fitness vs. Fatness on All-Cause Mortality: A Meta-analysis," *Progress in Cardiovascular Diseases* 56(4) (2014): 382–90.

29. Ming Wei, James B. Kampert, Carolyn E. Barlow, Milton Z. Nichaman, Larry W. Gibbons, Ralph S. Paffenbarger Jr., and Steven N. Blair, "Relationship Between Low Cardiorespiratory Fitness and Mortality in Normal-Weight, Overweight, and Obese Men," *Journal of the American Medical Association* 282(16) (1999): 1547–53.

30. Blair, "Physical Inactivity."

31. Xiao-Ren Pan, Guang-Wei Li, Ying-Hua Hu, Ji-Xing Wang, Wen-Ying Yang, Zuo-Xin An, Ze-Xi Hu, et al., "Effects of Diet and Exercise in Preventing NIDDM in People with Impaired Glucose Tolerance: The Da Qing IGT and Diabetes Study," *Diabetes Care* 20(4) (1997): 537–44.

32. Diana Thomas, Elizabeth J. Elliott, and Geraldine A. Naughton, "Exercise for Type 2 Diabetes Mellitus," *Cochrane Database of Systematic Reviews* (3): CD002968 (2006).

33. Kelly A. Shaw, Hanni C. Gennat, Peter O'Rourke, and Chris Del Mar, "Exercise for Overweight or Obesity," *Cochrane Database of Systematic Reviews* (4): CD003817 (2006).

34. Caroline K. Kramer, Bernard Zinman, and Ravi Retnakaran, "Are Metaboli-

cally Healthy Overweight and Obesity Benign Conditions? A Systematic Review and Meta-analysis," *Annals of Internal Medicine* 159(11) (2013): 758–69.

35. Margaret F. Gregor and Gökhan S. Hotamisligil, "Inflammatory Mechanisms in Obesity," *Annual Review of Immunology* 29 (2011): 415–45.

36. Samuel Klein, Luigi Fontana, V. Leroy Young, Andrew R. Coggan, Charles Kilo, Bruce W. Patterson, and B. Selma Mohammed, "Absence of an Effect of Liposuction on Insulin Action and Risk Factors for Coronary Heart Disease," *New England Journal of Medicine* 350(25) (2004): 2549–57.

37. Kai Sun, Christine M. Kusminski, and Philipp E. Scherer, "Adipose Tissue Remodeling and Obesity," *Journal of Clinical Investigation* 121(6) (2011): 2094–2101.

38. Tracey M McLaughlin, T. Liu, Gail Yee, F. Abbasi, C. Lamendola, G. M. Reaven, P. Tsao, et al., "Pioglitazone Increases the Proportion of Small Cells in Human Abdominal Subcutaneous Adipose Tissue," *Obesity* 18(5) (2010): 926–31.

39. Qiqi Lu, Mingming Li, Yu Zou, and Tong Cao, "Induction of Adipocyte Hyperplasia in Subcutaneous Fat Depot Alleviated Type 2 Diabetes Symptoms in Obese Mice," *Obesity* 22(7) (2014): 1623–31.

40. Julianne Holt-Lunstad, Timothy B. Smith, and J. Bradley Layton, "Social Relationships and Mortality Risk: A Meta-analytic Review," *PLoS Medicine* 7 (2010): e1000316.

41. Lisa M. Jaremka, Christopher P. Fagundes, Juan Peng, Jeanette M. Bennett, Ronald Glaser, William B. Malarkey, and Janice K. Kiecolt-Glaser, "Loneliness Promotes Inflammation During Acute Stress," *Psychological Science* 24(7) (2013): 1089–97.

42. Andrew Steptoe, Natalie Owen, Sabine R. Kunz-Ebrecht, and Lena Brydon, "Loneliness and Neuroendocrine, Cardiovascular, and Inflammatory Stress Responses in Middle-Aged Men and Women," *Psychoneuroendocrinology* 29(5) (2004): 593–611.

43. Kate Karelina, Greg J. Norman, Ning Zhang, John S. Morris, Haiyan Peng, and A. Courtney DeVries, "Social Isolation Alters Neuroinflammatory Response to Stroke," *Proceedings of the National Academy of Sciences, USA* 106(14) (2009): 5895–5900.

44. Bernardo Léo Wajchenberg, "Subcutaneous and Visceral Adipose Tissue: Their Relation to the Metabolic Syndrome," *Endocrine Reviews* 21(6) (2000): 697–738.

45. Marie-Christine Pouliot, Jean-Pierre Després, André Nadeau, Sital Moorjani, Denis Prud'Homme, Paul J. Lupien, Angelo Tremblay, and Claude Bouchard, "Visceral Obesity in Men: Associations with Glucose Tolerance, Plasma Insulin and Lipoprotein Levels," *Diabetes* 41(7) (1992): 826–34.

46. De-xiu Bu, Gabriel Griffin, and Andrew H. Lichtman, "Mechanisms for the Anti-inflammatory Effects of Statins," *Current Opinion in Lipidology* 22(3) (2011): 165–70.

47. Amandine Chaix, Amir Zarrinpar, Phuong Miu, and Satchidananda Panda, "Time-Restricted Feeding Is a Preventative and Therapeutic Intervention Against Diverse Nutritional Challenges," *Cell Metabolism* 20(6) (2014): 991–1005.

48. Kim S. Stote, David J. Baer, Karen Spears, David R. Paul, G. Keith Harris, William V. Rumpler, Pilar Strycula, et al., "A Controlled Trial of Reduced Meal Frequency Without Caloric Restriction in Healthy, Normal-Weight, Middle-Aged Adults," *American Journal of Clinical Nutrition* 85(4) (2007): 981–88.

49. Daniela Jakubowicz, Maayan Barnea, Julio Wainstein, and Oren Froy, "High Caloric Intake at Breakfast vs. Dinner Differentially Influences Weight Loss of Overweight and Obese Women," *Obesity* 21(12) (2013): 2504–12.

50. Eric M. Matheson, Dana E. King, and Charles J. Everett, "Healthy Lifestyle Habits and Mortality in Overweight and Obese Individuals," *Journal of the American Board of Family Medicine* 25(1) (2012): 9–15.

51. Kim T. B. Knoops, Lisette C. P. G. M. de Groot, Daan Kromhout, Anne-Elisabeth Perrin, Olga Moreiras-Varela, Alesssandro Menotti, and Wija A. van Staveren, "Mediterranean Diet, Lifestyle Factors, and 10-Year Mortality in Elderly European Men and Women: The HALE Project," *Journal of the American Medical Association* 292(12) (2004): 1433–39.

52. A. Janet Tomiyama, Britt Ahlstrom, and Traci Mann, "Long-Term Effects of Dieting: Is Weight Loss Related to Health?" *Social and Personality Psychology Compass* 7(12) (2013): 861–77.

53. Linda Bacon, Judith S. Stern, Marta D. Van Loan, and Nancy L. Keim,

"Size Acceptance and Intuitive Eating Improve Health for Obese, Female Chronic Dieters," *Journal of the American Dietetic Association* 105(6) (2005): 929–36.

54. Greer Hawley, Caroline Horwath, Andrew Gray, Alison Bradshaw, Lisa Katzer, Janine Joyce, and Sue O'Brien, "Sustainability of Health and Lifestyle Improvements Following a Non-dieting Randomised Trial in Overweight Women," *Preventive Medicine* 47(6) (2008): 593–99.

Chapter 14: Change Your Lifestyle, Change Your Health

1. U.S. Department of Health and Human Services, "2008 Physical Activity Guidelines for Americans."
2. Satoru Kodama, Kazumi Saito, Shiro Tanaka, Miho Maki, Yoko Yachi, Mihoko Asumi, Ayumi Sugawara, et al., "Cardiorespiratory Fitness as a Quantitative Predictor of All-Cause Mortality and Cardiovascular Events in Healthy Men and Women: A Meta-analysis," *Journal of the American Medical Association* 301(19) (2009): 2024–35.
3. Suvi Rovio, Ingemar Kåreholt, Eeva-Liisa Helkala, Matti Viitanen, Bengt Winblad, Jaakko Tuomilehto, Hilkka Soininen, et al., "Leisure-Time Physical Activity at Midlife and the Risk of Dementia and Alzheimer's Disease," *Lancet Neurology* 4(11) (2005): 705–11.
4. J. E. Blundell, C. Gibbons, P. Caudwell, G. Finlayson, and M. Hopkins, "Appetite Control and Energy Balance: Impact of Exercise," supplement, *Obesity Reviews* 16(S1) (2015): 67–76.
5. Ming Wei, James B. Kampert, Carolyn E. Barlow, Milton Z. Nichaman, Larry W. Gibbons, Ralph S. Paffenbarger Jr., and Steven N. Blair, "Relationship Between Low Cardiorespiratory Fitness and Mortality in Normal-Weight, Overweight, and Obese Men," *Journal of the American Medical Association* 282(16) (1999): 1547–53.
6. Fit Fatties Forum, fitfatties.ning.com/.
7. Holly S. Kessler, Susan B. Sisson, and Kevin R. Short, "The Potential for High-Intensity Interval Training to Reduce Cardiometabolic Disease Risk," *Sports Medicine* 42(6) (2012): 489–509.

8. Josephine Y. Chau, Anne C. Grunseit, Tien Chey, Emmanuel Stamatakis, Wendy J. Brown, Charles E. Matthews, Adrian E. Bauman, and Hidde P. van der Ploeg, "Daily Sitting Time and All-Cause Mortality: A Meta-analysis," *PLoS ONE* 8(11) (2013): e80000.

9. Genevieve N. Healy, Charles E. Matthews, David W. Dunstan, Elisabeth A. H. Winkler, and Neville Owen, "Sedentary Time and Cardio-metabolic Biomarkers in US Adults: NHANES 2003–06," *European Heart Journal* 32(5) (2011): 590–97.

10. David W. Dunstan, Bronwyn A. Kingwell, Robyn Larsen, Genevieve N. Healy, Ester Cerin, Marc T. Hamilton, Jonathan E. Shaw, et al., "Breaking Up Prolonged Sitting Reduces Postprandial Glucose and Insulin Responses," *Diabetes Care* 35(5) (2012): 976–83.

11. Gary Charness and Uri Gneezy, "Incentives to Exercise," *Econometrica* 77(3) (2009): 909–31.

12. D. L. Katz and S. Meller, "Can We Say What Diet Is Best for Health?" *Annual Review of Public Health* 35 (2014): 83–103.

13. Centers for Disease Control and Prevention, "State Indicator Report on Fruits and Vegetables," 2013.

14. Oyinlola Oyebode, Vanessa Gordon-Dseagu, Alice Walker, and Jennifer S. Mindell, "Fruit and Vegetable Consumption and All-Cause Cancer and CVD Mortality: Analysis of Health Survey for England Data," *Journal of Epidemiology & Community Health* 68 (2014): 856–62.

15. David J. A. Jenkins, Cyril W. C. Kendall, Augustine Marchie, Dorothea A. Faulkner, Julia M. W. Wong, Russell de Souza, Azadeh Emam, et al., "Effects of a Dietary Portfolio of Cholesterol-Lowering Foods Versus Lovastatin on Serum Lipids and C-Reactive Protein," *Journal of the American Medical Association* 290(4) (2003): 502–10.

16. Manisha Chandalia, Abhimanyu Garg, Dieter Lutjohann, Klaus von Bergmann, Scott M. Grundy, and Linda J. Brinkley, "Beneficial Effects of High Dietary Fiber Intake in Patients with Type 2 Diabetes Mellitus," *New England Journal of Medicine* 342(19) (2000): 1392–98.

17. Elinor Ochs and Tamar Kremer-Sadik, eds., *Fast-Forward Family: Home, Work, and Relationships in Middle-Class America* (Berkeley: University of California Press, 2013), 56.

18. Dariush Mozaffarian, Renata Micha, and Sarah Wallace, "Effects on Coronary Heart Disease of Increasing Polyunsaturated Fat in Place of Saturated Fat: A Systematic Review and Meta-analysis of Randomized Controlled Trials," *PLoS Medicine* 7(3) (2010): e1000252.

19. L. Hooper, C. D. Summerbell, R. Thompson, D. Sills, F. G. Roberts, H. J. Moore, and G. Davey Smith, "Reduced or Modified Dietary Fat for Preventing Cardiovascular Disease," *Cochrane Database of Systematic Reviews* (5) (2012): CD002137.

20. Maria Luz Fernandez, "Dietary Cholesterol Provided by Eggs and Plasma Lipoproteins in Healthy Populations," *Current Opinion in Clinical Nutrition & Metabolic Care* 9(1) (2006): 8–12.

21. Alain J. Nordmann, Katja Suter-Zimmermann, Heiner C. Bucher, Iris Shai, Katherine R. Tuttle, Ramon Estruch, and Matthias Briel, "Meta-analysis Comparing Mediterranean to Low-Fat Diets for Modification of Cardiovascular Risk Factors," *American Journal of Medicine* 124(9) (2011): 841–851.e2.

22. Lluís Serra-Majem, Blanca Roman, and Ramón Estruch, "Scientific Evidence of Interventions Using the Mediterranean Diet: A Systematic Review," supplement, *Nutrition Reviews* 64(1) (2006): S27–47.

Chapter 15: Good Habits Beat Good Intentions

1. Mindy F. Ji and Wendy Wood, "Purchase and Consumption Habits: Not Necessarily What You Intend," *Journal of Consumer Psychology* 17(4) (2007): 261–76.

2. Thomas L. Webb and Paschal Sheeran, "Does Changing Behavioral Intentions Engender Behavior Change? A Meta-Analysis of the Experimental Evidence," *Psychological Bulletin* 132 (2006): 249–68.

3. Roy F. Baumeister and Todd F. Heatherton, "Self-Regulation Failure: An Overview," *Psychological Inquiry* 7(1) (1996): 1–15.

4. Earl K. Miller and Jonathan D. Cohen, "An Integrative Theory of Prefrontal Cortex Function," *Annual Review of Neuroscience* 24 (2001): 167–202.

5. Wendy Wood, Jeffrey M. Quinn, and Deborah A. Kashy, "Habits in Every-

day Life: Thought, Emotion, and Action," *Journal of Personality and Social Psychology* 83(6) (2002): 1281–97.

6. Christopher J. Armitage and Mark Conner, "Efficacy of the Theory of Planned Behaviour: A Meta-analytic Review," *British Journal of Social Psychology* 40(4) (2001): 471–99.

7. Benjamin Gardner, Gert-Jan de Bruijn, and Philippa Lally, "A Systematic Review and Meta-analysis of Applications of the Self-Report Habit Index to Nutrition and Physical Activity Behaviours," *Annals of Behavioural Medicine* 42(2) (2011): 174–87.

8. Phillippa Lally, Cornelia H. M. van Jaarsveld, Henry W. W. Potts, and Jan Wardle, "How Are Habits Formed: Modelling Habit Formation in the Real World," *European Journal of Social Psychology* 40(6) (2010): 998–1009.

9. Phillippa Lally and Benjamin Gardner, "Promoting Habit Formation," supplement, *Health Psychology Review* 7(1) (2013): S137–S158.

10. Peter M. Gollwitzer and Paschal Sheeran, "Implementation Intentions and Goal Achievement: A Meta-analysis of Effects and Processes," *Advances in Experimental and Social Psychology* 38 (2006): 69–119.

Index